SPACE TOYS

SPACE TOYS
A COLLECTOR'S GUIDE
TO SCIENCE FICTION
AND ASTRONAUTICAL
TOYS
By
CRYSTAL AND
LELAND PAYTON

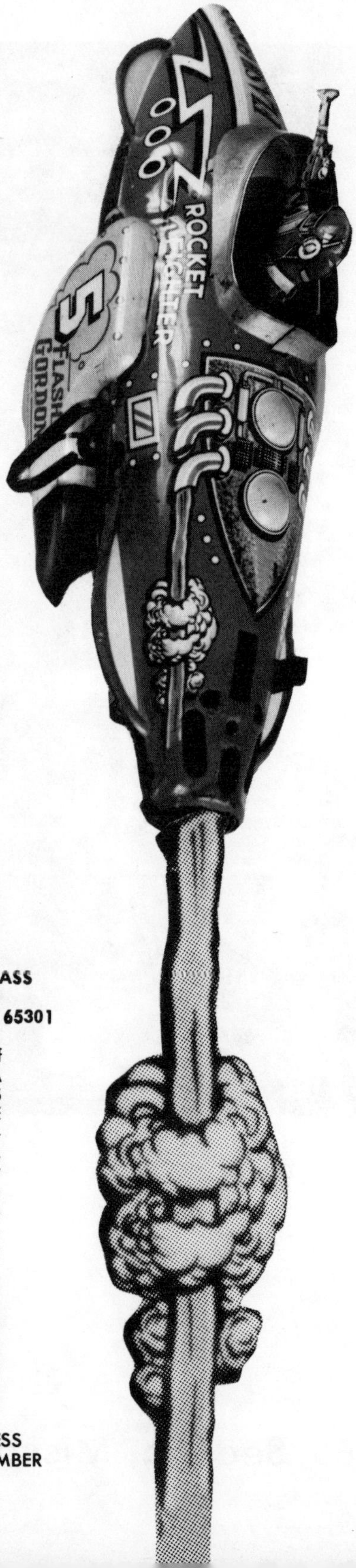

CONTENTS

Dedicated to the memory of
Crystal's great-uncle Jack,
Major General John F. Curry, U.S.A.F.

COLLECTORS COMPASS
P.O. BOX 1666
SEDALIA, MISSOURI 65301

 Printed in the U.S.A. (**ISBN 0-943464-00-5. FIRST EDITION.**)

LIBRARY OF CONGRESS CATALOG CARD NUMBER 82-73330

To understand space toys we must relate them to science fiction, popular adaptations of science fiction ideas, actual events concerning man's conquest of space and public perceptions of all of this. The speculations and conclusions we've made may disagree with yours and we may be wrong in fact. The book was done to throw light on the subject and hopefully to promote collecting these trifles which relate, as we said, to a great deal more. With light comes heat so we accept and welcome criticism or corrections. Let us know what you like or don't like and we'll take that into account in the next edition.

SPACE

Courtesy N.A.S.A.

ARMILLARY SPHERE. Until the Copernican Revolution the earth was believed to be at the center of the universe. These earth-centric ideas were quaintly represented in models called armillary spheres. The shock of realizing that the earth revolved around the sun was enormous. New discoveries about space will always have profound effects on all of us.

TOY SPACEMAN AND ASTRONAUT RUSSELL L. SCHWEICKART. The all plastic wind up American space toy is from the early 1950's. Astronaut Schweickart climbs about on the Lunar Module spacecraft during docking exercises on the Apollo 9 flight of 1969. Naive juvenile science fiction waned in the late 1950's as the space program gained momentum. When manned space flights declined in the late 1970's STAR WARS became an unexpected box office hit. To understand science fiction toys we must recognize the effect of the interwoven history of real and imaginary space travel.

Most of human history has been acted out within a thin film of gases that cling to the earth. We breathe this atmosphere using the oxygen; plants extract the carbon dioxide. Even mountain climbers encounter difficulty breathing after going up a few miles. At 16,000 feet half the atmosphere lies below. At 29,000 feet, the height of Mount Everest, 3/4ths of the earth's atmosphere has been exhausted. Above that gas molecules become more and more scarce finally disappearing almost completely. Through the blackness travel deadly radiation waves blocked from us by the mantel of the earth's atmosphere. Most meteors burn up in the atmosphere before reaching earth. Unprotected man would instantly freeze and explode in this environment.

True space begins in a region between 600 and 2,000 miles above the earth. There lighter gas molecules escape the gravity of earth and fly off into outer space. But at about 100 miles the laws of aerodynamics become unworkable and astrodynamics is needed if we are to travel.

Weightless astronauts float in space while traveling at speeds incomprehensible on earth. Yet the public seemed strangely prepared. There was interest to be sure but the conquest of space was made without disbelief or hysteria. Why? The answer is that they had been there before - in the comics, movies and short stories of science fiction. In fact, if anything, the real thing was anticlimactic. Jaunty, all-American boys that our astronauts were, they didn't encounter Ming the Merciless, Mook the Moon Man or even Killer Kane. There was a temporary deflation of pop science fiction about the time of the first space flights but the recent success of STAR WARS has proven that the need for drama beyond the earth is still with us. There is a psychic aspect to space and space affairs that must concern us if we are to understand the origins and history of space toys. In this "inner space" dimension of outer space deeply rooted stereotypes or archtypes recur time after time. One might be tempted to relate these recurrent types and familiar dramatic structures back to ancient Greek myths. Yet although sometimes rather naively executed such popular space stories have an element that qualifies them as legitimate science fiction.

They take place against a background that never existed, an alternate world created by the author by projecting technological changes on the real world. This is a relatively new theme in literature, one that required an awareness of the fact that science and technology could change our world. It wasn't until the Industrial Revolution in the late 19th and early 20th centuries that rapid and dramatic changes in society due to technological progress could be observed and commented on. Before that technology changed society but over the course of many generations. Such changes were too slow and diffuse to observe and to be attributable. Several important science fiction writers believe that FRANKENSTEIN, written in 1818 by Mary Shelley was the first true science fiction work. It was not concerned with space but it projected the effects of science upon human behavior imagining the consequences before the technology had been developed. Today with the extensive use of organ transplants, artificial body parts, genetic engineering and cloning, the Frankenstein saga has become a vital and timely theme - i.e. what will all this new power over life create, a superior human or a monster? In the same way BUCK ROGERS prepared us for space travel, perhaps even over-prepared us. The point is that science fiction anticipates the impact of technology and

APOLLO 14 LIFT OFF AND TIN TOY ROCKET BY SAN. Man may have dreamed of reaching the moon for thousands of years but for several hundred years men of science have realized that some day it would be achieved. Fictional accounts and toys acquaint the public with the look of successive technologies that might be used in space travel. This tin toy from the 1950's is modeled after the German V-2 rocket of the 1940's. It may appear ridiculous its portholes but it did acquaint the children who played with it with vertical lift offs — something the BUCK ROGERS toys didn't do — they took off horizontally, like airplanes.

gives us some chance to adjust to change before it occurs. Space toys, first a spin off of pop science fiction, then inspired by actual events in space, make those who play with them participants in the drama itself. They are typical toys in many senses, subject to the rules of any childhood possession but their referral to the drama of space cannot help but affect those who play with them.

What are these origins as they affect the popular perceptions of space? Space has been a convenient dramatic backdrop for over one hundred years. Jules Verne in FROM THE EARTH TO THE MOON (1865) described the weightlessness of travelers in a huge aluminum bullet. As science revealed facts about the region beyond the earth fiction writers incorporated these facts or speculated on human dramas acted out within these alien environments. The remoteness of space has also attracted writers. Poetic license can operate in the far corners of the universe unencumbered by the constraints of what we have experienced first hand. Space could be whatever you want it to be, providing every kind of conceivable alien villain or friendly robot as the dramatic structure dictated. After the astronauts penetrated space and especially after the Americans walked on the moon some of the more outlandish inventions about space came to a temporary halt. Gradually the writers learned to move all their pop space dramas into distant galaxies where man has never gone (and probably never will) or into the future or past using a time travel device. STAR WARS you will recall opens with lines rolling on the blackness of space: "A long time ago, in a galaxy far, far away..." Scientific truth is today so difficult for the layman to understand (and so subject to rapid evolution, as is technology) that a great deal of media science fiction avoids the burden of realistic style and concentrates on escapism, horror, the juvenile market, or sometimes a combination of all three. In an era of declining mass competence in science and waning belief in rationality, this is doubly understandable. The fate of space exploration itself may be in doubt given the temper of these times. Beyond the military and economic justifications there is a strong element of pure scientific curiosity in these heroic efforts to take man to the stars. Hopefully the mundane problems of today will not sidetrack us from the limitless opportunities which the exploration of the universe presents us with. We like to think that space toys themselves play some role in advancing the actual exploration of space by familiarizing children with those ideas. Some of the great classic Japanese space toys and robots were not commercial successes either. Their scarcity attests to that. Truth is, science fiction and science fiction toys like the space effort itself are sometimes better than they need to be.

SPACE AS A BLACK CANVAS. Space is, for writers, filmmakers and toy designers, a black canvas to decorate as they wish. The results often as not reveal more about the "inner space" of human psychology than the outer space of the universe. Awareness of these subjective and objective dimensions is required if we are to appreciate the importance of this vast backdrop in human affairs. Today's collectors cherish toys (left to right) such as the MECHANICAL INTERPLANETARY EXPLORER by AN, RENDEZVOUS 7.8 by Yanoman, and SPARKY ROBOT by KO. An interesting addition to such classic toys would be magazine and paper material that treat the space theme.

CLICK
THE NATIONAL PICTURE MONTHLY • FEBRUARY 1939
RADIO HYPNOSIS: The "MARTIAN TERROR" Explains What Is Happening to the GERMAN PEOPLE
A SIGNAL FROM MARS.
MARCH AND TWO STEP
RAYMOND TAYLOR
Arranged by
E.T. PAULL
SIMPLIFIED 4
FOUR HANDS 10
PUBLISHED BY E.T. PAULL MUSIC CO.
NEW YORK.
WEST 28th ST.
SWITCH
8
FLASH GORDON
FLASH GORDON
KING COMICS
December
AMAZING STORIES
25 Cents
HUGO GERNSBACK EDITOR
Miles J Breuer M.D
Clare Winger Harris
Scientifiction

KIDS AND COLLECTORS

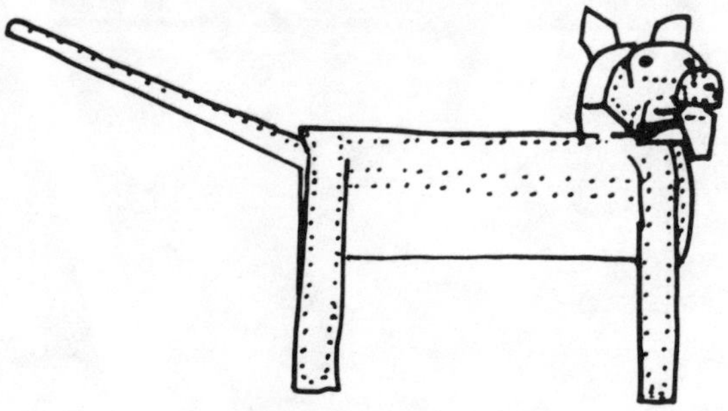

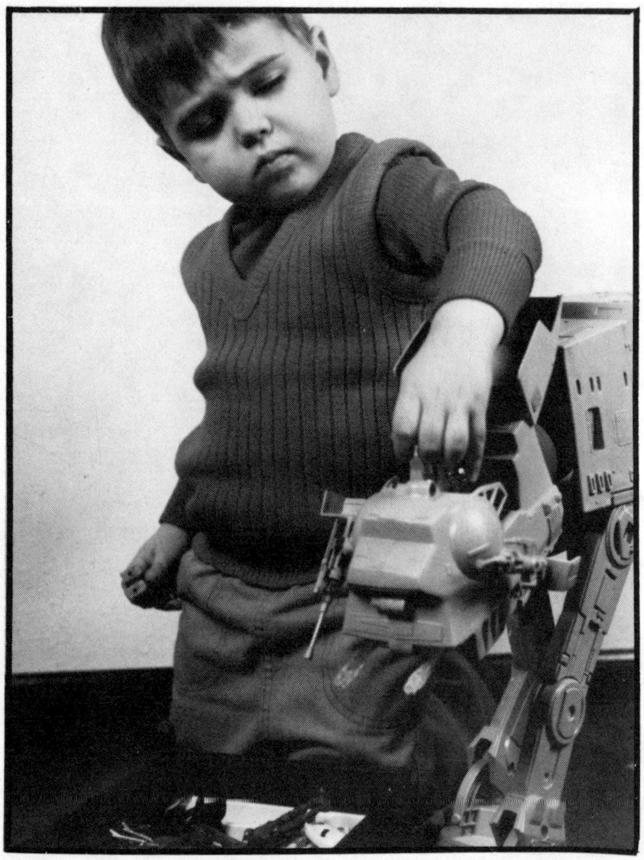

FOLK ART ROBOT DOG. Robot collector, Larry Whiteley, added studs to a naive dog sculpture and shows us the strange affinity of space art and primitive art. That first rate artists, designers and collectors find space toys saveable will have profound (and unexpected) influences on the development of this new collecting interest.

ALL TERRAIN ARMORED TRANSPORT (AT-AT). This Kenner-made 17½" tall replica of the Imperial Storm Troopers' 50 foot high attack vehicle is from George Lucas' film THE EMPIRE STRIKES BACK.

The full effects of mankind's exploration of the universe beyond the planet Earth are probably unrecognized by many of us. Living as we do, besieged by traditional problems of scarcities and cultural conflicts, the hi-tech dramas of outer space often seem unreal. In our age there is a widespread revulsion against advanced technology as witness the protests against nuclear power and the association some make between industrialism and pollution. Space exploration is less dramatically opposed but all of us have heard populist politicians voice concern that money should be spent on poverty programs rather than space programs.

Children and adolescents are spared many of the practical constraints of adults. Space travel is nothing but wonderful to them as witness the commercial success of STAR WARS and before that BUCK ROGERS and FLASH GORDON. Our 4 year old son, Strader, is only vaguely concerned with the political affiliation of his 50 odd (and some of them are odd indeed) STAR WARS figures. He recognizes that the Lucas cosmic drama involves good and bad guys but the gray grandeur of the Kenner manufactured AT-AT Snow Walker overcomes its use to transport the fascist storm troopers of the Empire. The $47.97 plus shipping and tax price tag of the Snow Walker is as unimportant to him as the billion dollar price tags of the moon shots. Ross, his 2 year old brother, included the words "robot" and "space" among his first 20 words. In no small way Strader's and Ross' obsession with space toys has prompted Crystal and me to do this book. This is not to underrate the ability of toy collectors to suspend disbelief and enjoy their space things with parallel zest and enthusiasm. In a world some feel has gloomy prospects space toys seem to have just the right balance of fantasy and realism. We believe in them and in both the kids and the collectors — the two of whom have fewer differences than one might imagine. The main differences may be that collectors have more pocket change and are larger.

Photographed by Kate Loye

SPACE CAPSULE. Ross admires the kinetic action of this S.H. space toy patterned after the Mercury/Gemini spacecraft of the 1960's.

L.D. WHITELEY WITH FRIENDS. Mr. Whiteley is a Los Angeles-based folk art dealer and robot collector. Larry put the rivets into a less than thrilling piece of folk art to create the perfect mechanical pet and the first known folk robot dog.

TOYS AS PLAYTHINGS: IMPORTANT TRIVIA

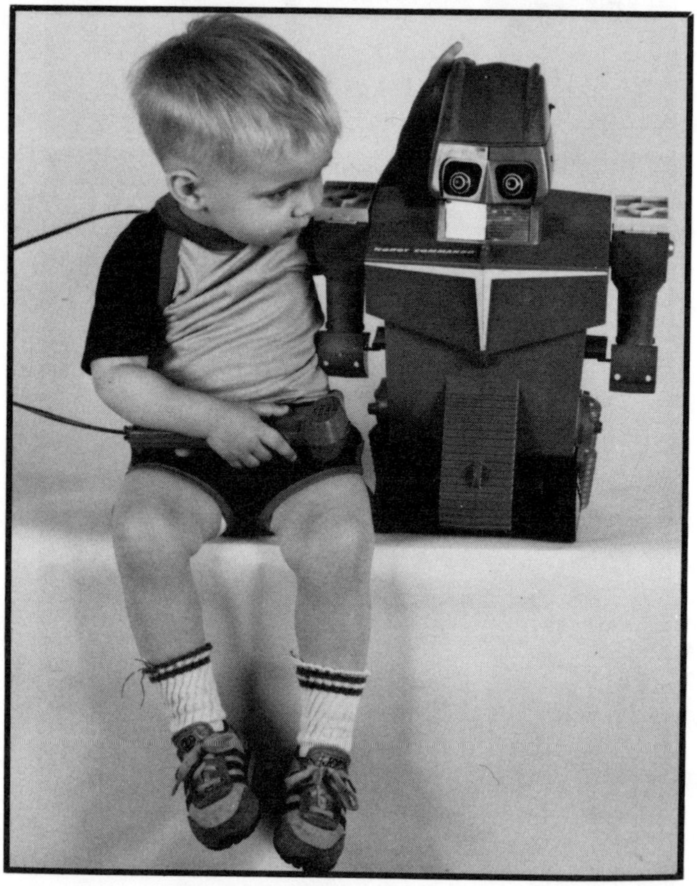

AN OLD TOY AS REMEMBRANCE OF THINGS PAST. In toy collecting as in psychiatry the role of childhood experiences affecting the unconscious is both undeniable and often equally unassessable.

ROBOT COMMANDO. Ross is fascinated by the big plastic robot made by Ideal. Most collectors aren't and such American toys go reasonably at toy shows and auctions.

"An object for children to play with." That seems fair enough for Webster's first definition of a toy. The dictionary's second definition, however, belittles that: "Something of little importance, a trifle." The real meaning of play and the importance of toys is a profound topic.

We rehearse for our roles in life in our play and our toys implement that play, helping to form our personalities. This process is complex. Obviously professions aren't usually created by toys or a third of the children of the turn of the century would have been transformed into firemen and another third would have become railroad engineers due to the popularity of cast iron fire and train toys. In the shadow of world economic and political reality, in the mix of parental aspirations for their children the economic needs of toy makers and the constraints of government regulation the toy is conceived. It is a thing born to be broken and used up, disfigured and discarded. Yet these "trifles" are our first experience with ownership, our first possessions — loved, fought over, and when lost, missed and cried after. All the problems of poverty —the joys and headaches are a part of the toy experience.

Consideration for the original use of the toy should be a conscious part of the collector's psyche. This is not to confuse nostalgia with the collector's larger esthetic and historical interests but rather to control sentimental impulses by understanding their roots in all of us. As the child is the father of the collector we cannot lose entirely the feelings of possessiveness children have for toys. The fantasy life of the juvenile does not disappear in adults. If you doubt these statements about the psyche of collectors attend a toy auction. Observe the anguish in the face of the bidders for a particularly rare old toy — then look at the rapture and inward bliss as the winner is handed his prize. To possess it he has thundered out with checkbook in hand to assert his ego. But once the toy has found a new home dreams are awakened and his thoughts soar above the haze of cigarettes and the din of other battles still going on. There has been a rekindling of the feelings of childhood — the fire of primitive possession and now the smokey flight of imagination.

But imagination can be the source of conjecture which ultimately produces logical ideas. If there are moments of childlike possessiveness in the heart of the toy collector they are certainly of smaller importance than the cerebral motivations, be they flights of fantasy or scholarly research. Toy collecting is getting to be serious business as well as a sophisticated hobby. Collectors of early German toys for instance have a 20 volume work to consult to make an authoritative decision. In such a world of objectification mature goals and rational methods will become even more useful. But collectors will always have motives that resist exposure. Without deeper meanings toy collecting wouldn't be so fascinating.

The study of toys and childhood memorabilia related to space has a considerable potential for complexity. As well as being regular toys they refer to and conjure up some of the most profound cultural, scientific, political and philosophical topics of our era. Space and robots involve conflicts and dilemmas of headline importance. It might be said that children (and some toy collectors perhaps) are oblivious to such things. Our observations are that no small part of the attraction our two boys have for space toys is due to their tie-ins with current events on the news, TV commercials (with their spacey backdrops and robot salesmen) and the general prominence of all media of the subject. All this play-up of space, robots, rockets and lasers creates prestige for their own small versions of them. Therefore we are going to deal with both the factual background of space exploration and popular culture's perception of it as influenced by science fiction and reconstituted (and frequently diluted) by comics, television and movies. Hopefully this will add a dimension to the collector's understanding and enjoyment of the toys.

STAR WARS MILLENNIUM FALCON. Toys can be a source of both joy and sorrow. Their role in developing personalities is possibly great but obscure. Similarly, adult collectors may be affected by childhood memories but caution must be exercised in making simplistic assumptions. One thing is certain — there is more than nostalgia and possessiveness operating at the high end of toy collecting where dedication and a willingness to absorb a lot of information are minimum requirements to play with the best old toys.

SPACE TOYS AS CLASSICS

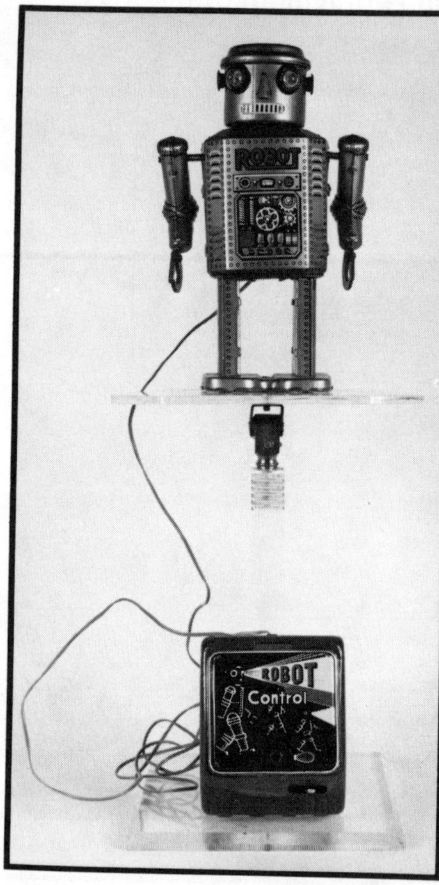

ROBOT AND AUTOMATON. Toys and figures that move in a human-like way have been amusing people for over 2,000 years. This 1980's plastic robot and the mandolin playing automaton of 1540 are intriguing footnotes in the history of automation whose main chapters are concerned with computers, guidance systems and robotics.

REMOTE CONTROL ROBOT. Reminiscent of a 1930's pulp cover this all tin robot was evidentally made by Masudaya Toys over a 10 year period. The earlier ones have a tin decorated battery box, the later ones a plain battery and control cylinder.

It may seem odd that recently made toys are sought after by collectors. That some have put 1950's robots beside their 18th century automata would appear to be incongruous. "Antique" and "collecting" seem to the public to be synonymous. It's no wonder. One hundred years ago "antique" meant "ancient". An antique was something made by the Greeks or Romans two thousand years ago. In the 1930's "antique" meant "made one hundred years ago." This definition came about for the convenient use of customs agents. As collectors of that era were interested almost exclusively in fine traditional hand made furniture and pre-industrial artifacts the 100-year-old definition of "antique" had some merit. But by 1980 there are few hand made items a hundred years old. 1880's stuff may look quaint but it's factory produced. To further assault the definition of antique a market developed for Art Deco, the decorative style of the 1920's and 1930's. "What to do°" said the promoters of antiques shows and the publishers of antiques magazines. "What do we allow, or include?"

Complicating matters, reproductions or copies of all kinds of old things began appearing and limited edition plates and figurines showed up, promising artificial scarcity as a substitute for the attrition of age. The confusion over "antique" remains due to the competing special interests of different coteries of collectors, dealers, promoters and publishers.

A better definition for antique than old is "obsolete." All of us know that styles change faster today than they did in the past. The effects of technology impact ever faster upon human affairs. Collectors are variously interested in curious things, things that are special, beautiful, unusual or just rare. Collectors are triggered into hoarding the object of their desires by scarcity. The way things move today, anything that stops being a shelf item is a legitimate antique we feel. If it turns someone on let him collect it with as much enthusiasm as those who save 18th century highboys. Those with a basement rec room of Billy Beer cans aren't likely to have the status of those with a drawing room of Chippendale's products but that's OK too. The point is that any artifact made by man is potentially deserving of being collected, loved, cherished, admired, studied and exhibited. Old shouldn't be the test, because everything will become that. Collectors are looking for special objects that fulfill some special criteria.

If that premise is accepted, then why not buy things new and wait until they become unavailable? It's a good idea for the seers and prophets among us. But few of us are so blessed with foresight that we can do that. Certainly a huge percentage of what is made achieves its highest desirability and price at that time. Collecting may become more specialized and esoteric but I predict that buying a K Mart and locking it up for 50 years will not capture many of the great collector's pieces of our times. Actually, had someone locked up a country store 50 or 100 years ago it might not have worked out as well as one might have expected. In spite of fabulous returns on certain knives, guns, toys and spool cabinets the vast amount of staples would bring little more today in constant dollar value and would drag the speculative effort down if even 2% interest on the original cost would be charged against the profits. The scatter gun approach to collecting/speculating is doomed to failure. Only a precious few things made by man are destined to be appreciated and to appreciate in value beyond their original retail cost. It takes knowledge and intuition to predict what shelf items might be valuable in the future and nerve to carry out the plan. All this came together in the person of L.D. Whiteley, Southern California toy collector and dealer.

The very first day we met Larry he showed us a box of Japanese made robots. Regretfully, his defense of the toys fell on deaf ears. At that time we were interested in folk art and scientific instruments. A number of the robots Larry said were then obtainable for $10 to $20 are $50 and $100 items today, 5 years later. Not only did we let Whiteley's reasoning drift by us, we sold him a robot dog for $25 two years later that we'd buy back for a hundred today ... maybe more. When we offered it to him for $10 and he paid us $25 then at least a small light should have gone on.

As a big time toy dealer, Larry had exposure to the classic criteria of a good collectible toy. Toy enthusiasts are, I think, unusually rigorous among collectors about their subject and have developed a logic that Larry had the brilliance to correctly apply to new objects. About the same time others worldwide seem to have arrived at the conclusion that robots and other contemporary toys relating to man's exploration of space were to be excellent perhaps even great collector's toys, but Larry's warehouse of robots attests to his acumen.

As a well established collector and dealer in classic antique toys Larry understood that a great toy has certain attributes and robots fulfilled these criteria. What is a legitimately collectible toy, or "legit" as Whiteley would say? First, it must be an industrially manufactured piece, not a product of an individual's effort like folk art. Folk art is a separate category (and one not necessarily abjurred by toy collectors - Whiteley collects and deals in folk art too). Good toys are unashamedly mass produced hence employ industrial technology current to their era of production. They frequently reflect the interests of their times in terms of subject. Toy tinplate trains were produced within a few years of the first trains. The 1909 catalogue of the great German toy maker, Marklin, illustrates the Wright biplane that flew at Kitty Hawk for the first time in 1903. Thus there is often a feeling of avant garde technology in both the subject and construction of a new toy that becomes with time a classic. The pages of important toy books are filled with illustrations of toys that although they appear old fashioned today, mirrored the latest fashions of the age in which they were made. In page after page of THE TOY COLLECTOR by Louis H. Hertz we see iron banks of 19th century politicians, scale locomotives, realistic bicycles and steam engines, countless fire rigs, boats and airplanes that capture the flavor of their period. There is a conspicuous absence of generalized toys. Good toys possess the ambience of their age. Construction methods and materials characteristic of that period are used frequently and colors in vogue are employed. Whether or not the toy is an exact scaled down copy doesn't matter as long as the spirit of the original is preserved. David Pressland notes in THE ART OF THE TIN TOY how the addition of a swan's-neck tiller to the Marklin 1909 Wright Brothers biplane adds charm to a very precise representation. Pressland's remarkable book, much of it in color, conveys the graphic effect of great toys - their capture of the flavor of the period - the perfection of their original finish - the glossy paint, the tiny pressed details, the neat little rubber tires. Their appeal is total.

The great classic toys are clever and well made. They were a material fulfillment of the optimism of the middle class for their children. All parents love their children and lavish gifts on them, but the middle class created by the Industrial Revolution expended unprecedented efforts in the technology of toys just as they did in the real world. Like the modern cities, bridges, railroads, and airplanes they brought into being, their classic toys are competent in execution and confident in spirit. Some believe their expressiveness is equal to that of painting and sculpture: i.e. THE ART OF THE TIN TOY.

Larry Whiteley often speaks of the sculptural quality of the toy robots he favors. This is certainly a similarity to the great toys of the past. Fine toys have amazing three dimensional qualities. They carve space decisively and survive both the distortions of photography and the perils of exhibition in large galleries with their original charm and verve intact. Other points of favorable comparison between space toys and great classic toys are authenticity to historic events, progressive use of materials and care of manufacture. Toys that use power, be it clockwork or steam, are especially cherished by children and collectors. Some of the battery operated bump-and-go space capsules and robots with their noisy, flashing colored lights and complex actions must be seen as the logical heirs to the action toy tradition. It must be recognized that all space toys aren't candidates as future classics but even many of these are redeemed by character tie-ins or sheer funky charm. For now let us concede that the best of the 1950-1975 Japanese made robots and space vehicles have sufficient merits to be collectible. Earlier we hopefully established that age alone does not guarantee value as a collectible. Many of the robots especially don't seem to have been made in large numbers and this combined with their inherent delicacy has early on produced some apparently rare pieces

and correspondingly steep prices for 7 to 30 year old trifles that sold originally for less than $10.

David Pressland concludes THE ART OF THE TIN TOY with a chapter titled "The Influence of Space and the Media 1945-1975." He admits that many of the post War tin toys are "uninspired" as are many older toys. "Yet," states Pressland, "The best of them readily stand comparison with their illustrious forebears." Louis H. Hertz in the last chapter of THE TOY COLLECTOR, entitled "Distinguished Past, Promising Future," mentions types of contemporary toys that may be desirable in the future. Hertz speculates toys of future value may include "Any toy based on contemporary personages, real or imaginary, but most particularly real people who are in the news. However, in general, this would include comic strip, television, and fictional characters." He continues his guess about collectibles of the future: "Any toy based on contemporary events, and most particularly at this writing, those based on space exploration and travel." The book was copyrighted in 1966. Fifteen years later his words have proven true. The day of the collectible space toy has arrived.

Toys have become blue chip collectibles lately. While we have seen certain rare space toys priced in the $2,000 to $3,000 range the most cash that has changed hands for one is around $1,500. Pamela Brown Sherer, head of the Collectibles division of Sotheby's, New York, informs us that the auction record of any toy was achieved at their December, 1979, New York sale when a 1905 Marklin painted tin clockwork battleship 34" long sold for $21,000. This famous auction house broke the $5,000 barrier for toys back in 1976 when they knocked down a Hubley cast iron four seat brake for $5,750. Since then a number of French 19th century automatons, really nice German boats and trains, and rare cast iron American mechanical banks have brought between $5,000 and $10,000. Ms. Sherer indicated they had not yet achieved stellar prices in the space toy category but that surprises in the collecting field are not unheard of. She related the extraordinary bidding for lot #782 of their December Collectors Carousel (a Marklin live-steam gauge O locomotive and tender). Estimated at $400 to $600 it sold for $6,500. "I hope that in the not too distant future we will see this level of prices for robots," she concluded. We agree that when the volatile new market matures there will be important space toys that may reach very surprising prices. As the chapter on "Space Toys as Classics" indicates some space toys are very much in the spirit of the fine old toys that are bringing spirited bidding and changing hands for serious money.

TOYS OF TRANSPORTATION. Toymakers have provided us with a record of our enthusiasm for speed. Tin toys have been especially suitable vehicles for expressing the technological advances that have taken us from horseback to space travel in 150 years. Now that plastic is the dominant material collector interest in all tin toys runs high. The best tin space toys seem destined to be classics having fulfilled many of the criteria set by the highly valued earlier ones.

SPACE TOYS AS MEMORABILIA

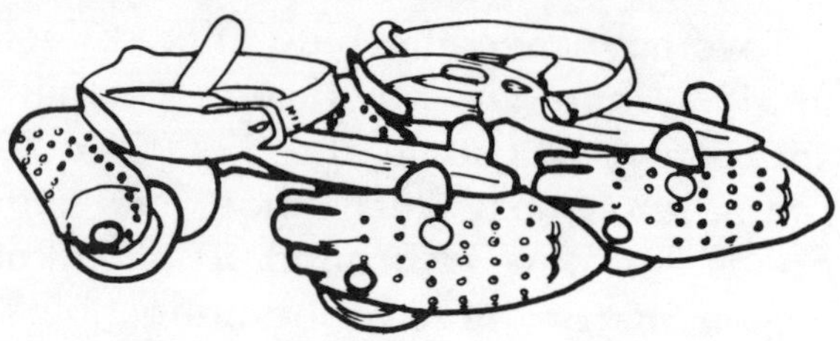

BUCK ROGERS' SKATES. Certain licensed BUCK ROGERS IN THE 25TH CENTURY toys from the 1930's can be wonderfully desirable and impossible to find today.

TV MEMORIES AS TOUCHSTONES. The specialness of toys, games and premiums from one generation's childhood may or may not be apparent to other generations. For those of us who grew up worrying that the picture tube would go on the blink before Captain Video could escape the clutches of Mook the Moon Man all things great or small, paper or plastic, imprinted with such heroes' names deserve rescue from the cluttered cases of flea market hustlers. After all, Tom Corbett, Captain Video and the TV Flash Gordon rescued us from the pervasive ordinariness of the Eisenhower years. Top to Bottom: TOM CORBETT POLARIS, tin key wind spaceship by Marx. CAPTAIN VIDEO SPACE GAME by Milton Bradley. CAPTAIN VIDEO ROCKET LAUNCHER by Lido Toy Co. FLASH GORDON SOLAR COMMANDO SET by Premier Products.

There are collectors who acquire toys based on their subject matter, character tie-ins or the nostalgia they evoke. Often as not they are unconcerned with their esthetic appeal in the formal sense that the collectors of classic toys are. Sometimes in fact the more blatant the violation of accepted good taste the more endearing the item is to them. Ephemera is to these collectors as important as a substantial piece with sound design, good materials and careful workmanship. Crudely lithographed paper give-aways or fragile inexpensive plastic premiums can command a place on their shelves. In defining and describing several character collectibles that relate to space we have made an attempt to convey the ambience of the fictional work. This can include some background photographs of the real world of that time. Some might focus on the sense of irony and absurdity of such juxtaposition but this isn't a put down.

We want to sympathetically portray this collecting interest and straight on museum photographs alone wouldn't do it. Memories are notoriously difficult to keep in focus. It must be admitted too that a sense of the ironic inhabits every such collector. True, some savers of Buck Rogers approach the subject with the dispassion of an archivist. At the other extreme there are those who pursue "camp" or "kitsch" - things selected for the outrageous irony they provoke. It's hard for anyone to be immune to the charm of an X2-38 Buck Rogers Disintegrator Pistol which could make buildings vanish by "an electronic barrage created by an electronic compression chamber and a sub-atomic condenser discharged by an impulse generator controlled by a beam intensity selector and aimed by a disintegration-beam director." Naturally to keep the barrel from melting the 50¢ weapon has "a tri-thermal convergence unit." But it takes more than a fleeting, wry smile to shell out $50 for one today. And what about Buck's roller skates, made by Marx, that look like spaceships and could be had from a dealer at the 1981 September Brimfield for a mere $2,000? Some serious love of the whimsical, huh, Space Rangers? The skates didn't roll off that day for two grand, incidentally. We have to admit that it's the only pair we've seen for sale.

Collectors should be aware however that recent space memorabilia particularly STAR TREK and STAR WARS paper items have been issued in tremendous quantities. It is not known in fact if some of this material will ever go out of print and appreciate in value. The manufacturers, distributors or hoarders of such items may attempt to manipulate the market much like the limited edition plate and figure producers suggesting that now is the time to buy before all two millon STAR TREK license plates disappear from the face of the Earth. These cult collectibles such as posters, jacket patches, buttons, gum cards, jewelry, T shirts, records, blueprints of spaceships and so called "collector's edition" magazines and books offer few manufacturing problems to keep them from perpetually being reissued in identical form to the original. They are marginal to the subject of this book so we will leave them untreated except to advise collectors to purchase such current items for their informational or esthetic appearance not because of a hope for monitary gain. (This does not refer necessarily to toys which do become unavailable in a season or two.) Some of this stuff may prove to be fabulously desirable but deprived of the knowledge of the size of editions and the intent of the publisher to keep reproducing the item it seems impossible to predict. Pre- 1970's space related memorabilia on the other hand is already difficult to retrieve. There were few collectors or dealers before then. We know of only a few reproduction Buck Rogers, Flash Gordon, or Captain Video premiums or ephemera but when prices rise enough one might beware of more facsimiles in the case of paper items or very simple plastic or metal things. There are various reissues however but they are not manufactured to fool collectors.

FLASH GORDON SET OF PLASTIC FIGURES. Each time a character is revived a new cycle of products is released. This set by Premier Products Co., Brooklyn, N.Y., dates from 1952.

ASTRONAUTICAL TOYS

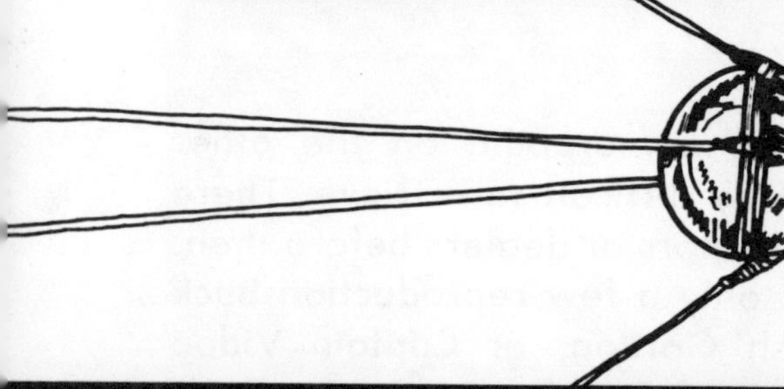

SPUTNIK 1. The world was shocked with the Soviet Union's launch of the first earth orbiting satellite in 1957. Toy makers were perplexed by the shape and few successful toys were made of the early satellites.

TOY SPACE CAPSULES AND 1960'S LIFE MAGAZINE. Not only did the Mercury and Gemini spacecraft offer toy makers a better design to copy they could put men inside them. The bottom two tin Japanese space capsules are battery operated. The top one is friction. Old LIFE's make an interesting adjunct to a space toy collection and one readily available for a few dollars or less.

In 1981 one of the Soviet astronauts participating in the Salyut 6 space program became the 100th space traveler. Twenty years before Major Yuri A. Gagarin of the Soviet Air Force became the first man in space. Four weeks after Gagarin's flight Alan B. Shepard, an American Naval officer, reached a height of 116 miles in a Mercury spacecraft test. The spacecraft actually used in this 20 year heroic era of the conquest of space did not resemble the sleek streamlined spaceships of Buck Rogers and his friends. Both the Soviet and American designs had a functional, matter-of-fact appearance. Soviet space toys were produced we understand but the Soviet embassy and trade delegation indicated it would be "impossible" to obtain pictures or examples of their astronautical toys.

The astronautical toys marketed in the U.S. that realistically resembled the real spacecraft of the 1960's and '70's were produced primarily in Japan. Some precisely copy the form of the orbiting capsules and the lunar module. They seem to have been marginally popular as toys. The bump and go battery operated realistic capsules and landers are quite attractive in action but they may be too delicate to play with. Collector interest in these small, realistic replicas is low at this time. Perhaps they do not fulfill that all important mythological component of science fiction as well as the fanciful imaginary spaceships of the Buck Rogers and Flash Gordon era. Interestingly reality has not caused toy makers to abandon the futuristic mode of spaceships. Spacecraft that Buck would feel at home in will probably be on the toy shelves for many years to come.

An interesting development in astronautical toys is the space shuttle which has considerably more streamlining than the earlier cones and modules. Graceful and swept-finned as it is toys modeled after the space shuttle, Columbia, must face stiff competition with the high tech futuristic, baroque designs originating in STAR WARS. Even so toys based literally on real spacecraft seem to us to be an interesting collecting category. Compared to robots they are quite abundant and affordable at present. Astronautical toys might be termed "Science Fact Toys." Although liberties are often taken with regard to exactness of replication of design they are clearly inspired by actual space exploration.

Battery operated or windup astronauts can be quite realistic or highly science fictionalized. They often seem to be more robot than human. They have all the merits of classic robots and attract the same audience. Several astronauts would bring over $500, mint in the box.

APOLLO 15 EVA AND TOY LUNAR MODULE BY DSK. Astronaut James Irwin, pilot of the lunar module, salutes Astronaut David Scott who records this EVA (extravehicular activity) of the 1971 mission to the moon. Strader Payton (who says he intends to be an astronaut when he grows up) is recorded beside his toy lunar module by his father.

Courtesy N.A.S.A.

TYPES OF SPACE AND SCIENCE FICTION TOYS

CLASSIC TIN JAPANESE TOYS C. 1950-1970. With bright lithographed thin steel bodies and astounding mechanical or electrical actions these toys are seen by some collectors as the rightful heirs of the great tin toy tradition. Seldom in the history of toy making has more ingenuity in design and competence in manufacturing been lavished on small, inexpensive trifles. Made to be played with by children of a people far across the ocean these toys at their best show both the artistry and technical genius the Japanese are capable of.

ASTRONAUTICAL REALISTIC. The battery operated EAGLE LUNAR MODULE by DSK, Japan, moves in an upright position. Not as big or detailed is Cragstan's friction powered FRIENDSHIP 7 SPACE CAPSULE. While these toys take liberties with scale and detail they bear a striking resemblence to N.A.S.A.'s real spacecraft of the 1960's and 1970's.

ASTRONAUTICAL FANCIFUL. From BUCK ROGERS to STAR WARS toy designers have been adapting the spacecraft designs of science fiction to toys. Illustrated is a BUCK ROGERS ROCKET by Tootsie Toy, next to a N.A.S.A. lift off with astronaut Scott Carpenter.

PLASTIC SPACE TOYS. Recent Japanese and Hong Kong toys and American made toys from the 1940's on have frequently been cast from plastic. Some of these, such as this BIG MAX AND HIS ELECTRONIC CONVEYOR, show interesting designs but at present they bring far less than the Japanese tin classics. Some all plastic space toys may be sleepers.

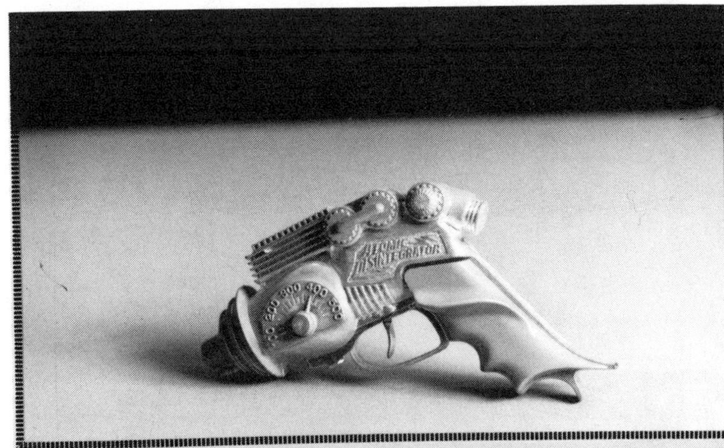

SPACE WEAPONS. One could collect nothing but toy space guns and fill a room. This ATOMIC DISINTEGRATOR by Hubley dates from the late 1940's.

MEDIA SCIENCE FICTION TOYS AND MEMORABILIA. Space and science fiction radio, TV and films have been the source of many toys and playthings. They range from the tenuous tie-ins of the LEAVE IT TO BEAVER ROCKET TO THE MOON space game to STAR WARS figures directly modeled after the most thoughtfully designed space opera of all times. Memorabilia collectors care more about the references to the fiction work than the intrinsic merit of the toy. Many valued pieces of memorabilia are of the cheaper materials produced with little manufacturing skills.

ASTRONAUTICAL MEMORABILIA. Playthings derived from real space events like this jigsaw puzzle with the pictures of astronauts Aldrin, Collins and Armstrong are not widely collected but it seems to us an interesting category.

NON SPECIFIC SPACE MEMORABILIA. (Left to Right) LUNCH PAIL by Thermos, LIFE magazine with astronaut on cover, MCDONALD'S HAPPY MEAL FLYING SAUCER container, ERECTOR ROCKET LAUNCHER SET, SATELLITE BANK by Duro Mold, and SNOOPY AS ASTRONAUT MUSIC BOX by Schmid Brothers. Memorabilia must compare favorably only with memories to be valued. While therefore less objectively related to the history of toys these diverse space related items may be seen as intriguing statements of the effects of the space age on popular culture. They are marginal as collectibles at present hence inexpensive and abundant.

ZAP-ZAP YOU'RE DEAD

SPACE GUN BY S.H., JAPAN. In the design of toy space weapons the bizarre is the rule. Whole collections could be made of futuristic space guns from Buck Rogers to the present.

SUPER ASTRONAUT BOX. Made by S.J.M., Taiwan, this contemporary robot has been in continual production in various forms in Japan for over 15 years. The earlier ones are virtually all tin but this one has a plastic head and arms. It walks then stops, the chest popping open and as the body rotates machine guns light up and loud shooting noises are guaranteed to command respect in the under-6 crowd. Toys of destructive potential are as fascinating to children as weapons are to most world leaders.

Without counting the numerous toy ray (and now laser) pistols it seems that half of all the space toys have some armaments. The little tin astronauts pack futuristic rifles or wear pistols - and they're not smiling. Space tanks fairly bristle with cannons. A lot of robots are peaceable, probably because of the serious implications of violent robots. Who wants to dwell on the idea of machines turning against their makers? Occasionally a robot has fire-power. The first time we tried out the "S.-J.M. Rotate-O-Matic Super Astronaut" the little caution, "Not recommended for children under 3 years of age," on the box took on a new meaning. Ross, then not quite two, generally loved the pretty lights and the bump-and-go action of the battery operated space toys. When the chest of the "Rotate-O-Matic" opened the stacatto machine gun bursts drowned out the gentle slosh of the dishwasher. Ross took one took and overpowered both with his cries of distress. Several weeks later we tried it again and this time he liked it. It is human nature to be afraid of and yet come to find terrible weapons irresistible.

Space was of course conquered by the coming together of two technological traditions, rocketry and computers. Rockets (aside from some peripheral use for amusement) were developed and continue to be developed as weapons. Most scientific probes into space leave earth on top of missiles originally designed to carry nuclear warheads. Computers, the control systems whose speed and automatic decision making capability, have as their ancestors simulacra and atomata as rich and varied as a 2,500 year old water powered alarm clock and a mechanical duck of 1738 that ate fish, wiggled its tail and excreted the remains. Of course, the parents of the modern computer are the calculating machines of

Charles Babbage (1791-1870) and the many innovators in mathematics and electronics of the last 100 years. The idea of machines with sophisticated automatic control systems (robots) is very old and not very military so we will for now return to rockets and futuristic weapons as they relate to space toys. Automata or robots have a rich if more peaceful history and deserve their own chapter - FACE OF THE MACHINE.

The intertwining of toys and other harmless devices for amusement with deadly serious artifacts is a recurrent theme in the conquest of space. We mentioned the linkage of computers and robots in sophisticated 18th century adult parlor toys. Rockets, according to some accounts, were first used for festivals and celebrations by the Chinese around 800 A.D. But by 1232 they had realized their destructive potential and "fire arrows," likely gunpowder rockets, were instrumental in defeating the Mongols who beseiged the town of Kai-fung-fu.

The Europeans learned of gunpowder and rockets through the Saracens and quickly began improving both for military use. The English monk, Roger Bacon, discovered that saltpeter added to black powder produced a faster, hotter burn. He wrote in code in 1249 "of saltpeter take 7 parts, of hazel branches 5, and like amount of sulphur and so with those call up thunder and destruction, if thou know the art." Albertus Magnus, the German scholar, published his formula for rocket propellant about the same time. In 1405 the German tactician, Konrad von Eichstadt, mentions three kinds of military rockets - vertically rising, floating rockets and rockets that run along wires. For some time rockets were more reliable weapons than the primitive guns or cannons. By the 16th century improvements in the gun and artillery put European military rocketry on

40 YEARS OF TOY SPACE WEAPONS TECHNOLOGY. Strader finds the BUCK ROGERS ATOMIC PISTOL by Daisy as potent against imaginary aliens as Ross' new STAR WARS LASER RIFLE by Kenner. Fantasy weapons know no obsolescence for their purpose is play, not war.

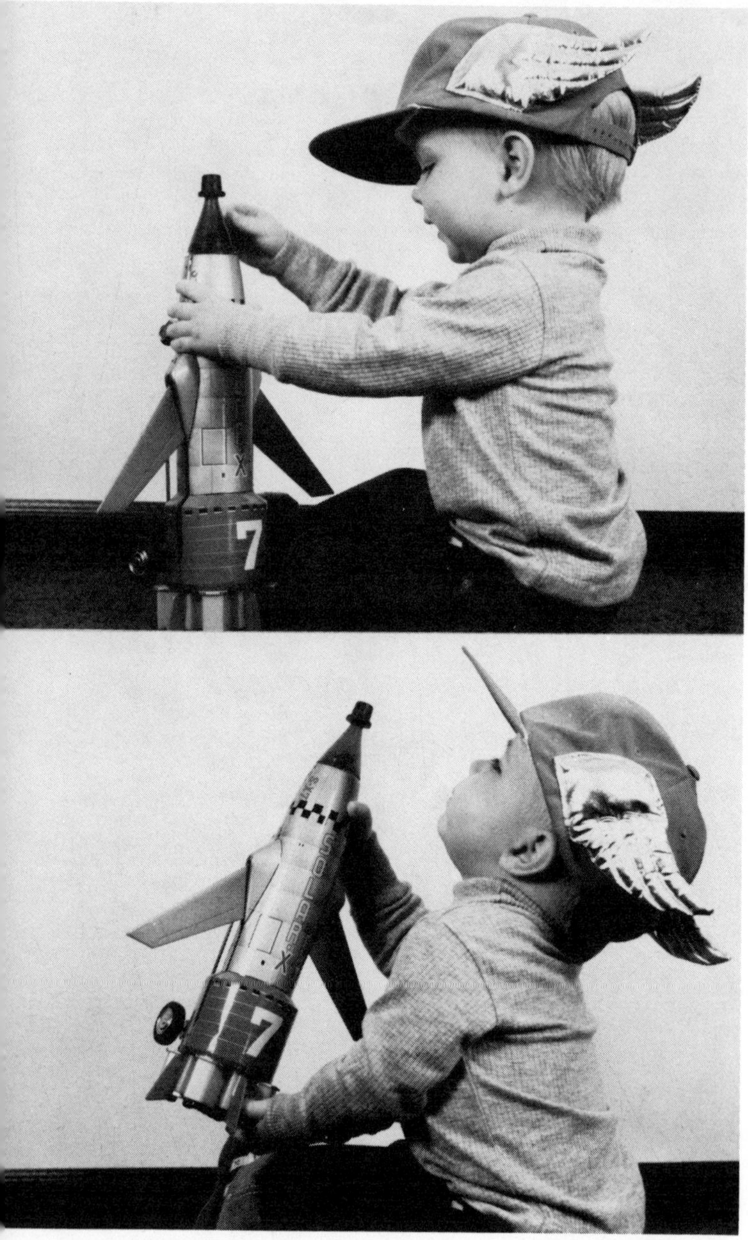

SOLAR-X SPACE ROCKET. The dream of flight likely existed before man invented the wheel. From the legend of Daedalus to the flights of our astronauts stories of soaring above the earth ignite our imaginations.

the back shelf - until rockets were used against the British in India with devastating results. In 1761 an Indian rocket corps of over 1,000 men fired on the British in the Battle of Panipat. The 10 foot long Indian rockets were cased in iron and could travel up to a mile and a half. Again in 1792 the British are rocketed in the Third Mysore War. Needless to say Indian rockets were captured and taken home where one Colonel William Congreve analyzed and began experimenting with them. With scientific methods he built a better engine of destruction and when the 32 pound Congreve rockets were first used in combat during the Napoleonic Wars they overshot their military target and set fire to a French town. We got a taste of Congreve's rockets in the War of 1812 when Fort McHenry in Baltimore was attacked. Francis Scott Key wrote THE STAR SPANGLED BANNER about this incident in a poem later set to music which asks if we can see the flag "by the rocket's red glare." Congreve, by the way, was by this time a knight honored for his contribution to improving the technology of weapons of mass destruction.

Does all this have a familiar ring? One country jolted into rocket research and development program by the success of a presumably backward nation - The capturing of enemy rockets to be analyzed and improved back home - The honoring of those who deliver the finest instrument of death using the latest methods of science? Writers as diverse and distant in time from each other as Francis Scott Key (THE STAR SPANGLED BANNER) and Norman Mailer (OF A FIRE ON THE MOON) find something dreaded but inspiring in the spectacle of burning rockets driving up through the air.

These long ago events seem to echo the U.S. space race, begun by the Soviet launch of the Sputnik satellite, the capture of the German V-2 rockets at Peenemunde and their transportation (along with many of their scientists) to White Sands, New Mexico to become the basis of our present rocket program, and the ultimate honors bestowed on Dr. von Braun and former German General Dornberger for their achievements.

If nationalist motives for ever more awesome weapons has been a theme in the development of the rockets that ultimately took man beyond the earth and into space there is a distinct counterpoint of pure scientific curiosity in the motives of those who actually made it all work. Although both the United States and the Soviet Union had important pioneers in rocket research their efforts were thwarted by a lack of adequate funding. The seminal rocket for both the present American and Russian space programs was the captured German A-4 rockets, called by us the V-2. In 1944, 4,320 V-2's equipped with one ton amatol explosive warheads were launched toward Britain in retaliation for the Allied bombing of German cities. The quick acceptance of the captured German scientists by the governments of both the U.S. and the U.S.S.R. has led to some cynicism about the nature of scientific responsibility. However it must be noted that the split up German rocket experts have cooperated fully with both countries, displaying apparently little interest in politics. Almost every scientist that worked on developing the V-2 came from the Verein fur Raumschiffahrt (The Society for Space Travel). Wernher von Braun joined in 1930 when he was but 18. As a private group they had little money for research. The German military had unlimited funds.

Hermann Oberth, a professor of physics and mathematics, was the guiding light of

Courtesy N.A.S.A.

APOLLO 11 LIFT OFF. For the first time in human history men will walk on the moon. The drama of the Apollo 11 mission will never be forgotten.

the society. As early as 1923 he published a book THE ROCKET INTO INTERPLANETARY SPACE. An interesting sidelight is that professor Oberth accepted the position as technical advisor of the Fritz Lang movie, FRAU IM MOND (WOMAN ON THE MOON). Von Braun, one of his proteges, was also to grasp the importance of media coverage of rockets and space both in fact and fiction. His credits run from books to television to designing toy model kits and amusement park space vehicles for Walt Disney. Willey Ley a charter member of the Society for Space Travel also designed model rockets and even consulted on TOM CORBETT, SPACE CADET, the early 1950's children's television show. For all the potential destructiveness of rockets it's hard to make bad guys out of scientists who take time to create toys and childhood fantasies. The idea of adventures in the limitlessness of space is a dream common to both the young and the visionary geniuses of scientific rocket research.

Perhaps there is nothing Faustian about children's lust for limitless power but the rocket people have certainly occasionally struck bargains with such dubious earthly types that one could easily imagine Goethe's hero today trading his soul for a government grant to build a nuclear powered starship. Time and again the paradox of our technical mastery of nature and our primitive capacity for destructiveness comes to us as we survey the shelves of our collection of space toys. In gaudy lithos on tin and molded in plastic are the artifacts of avant garde scientific ideas - ray guns and laser pistols, tanks that can crawl on Mars, rockets that promise to hurl us into the darkest recesses of outer space. It's some comfort to know that at the helm of most are steel jawed Buck Rogers, Flash Gordon or Luke Skywalker or similar if anonymous types. With their humorous side kicks, their rebellion against tyrants and their eye for pretty girls how can we be afraid of the fearsome powers that come from the barrel of their plastic rockets and tin guns?

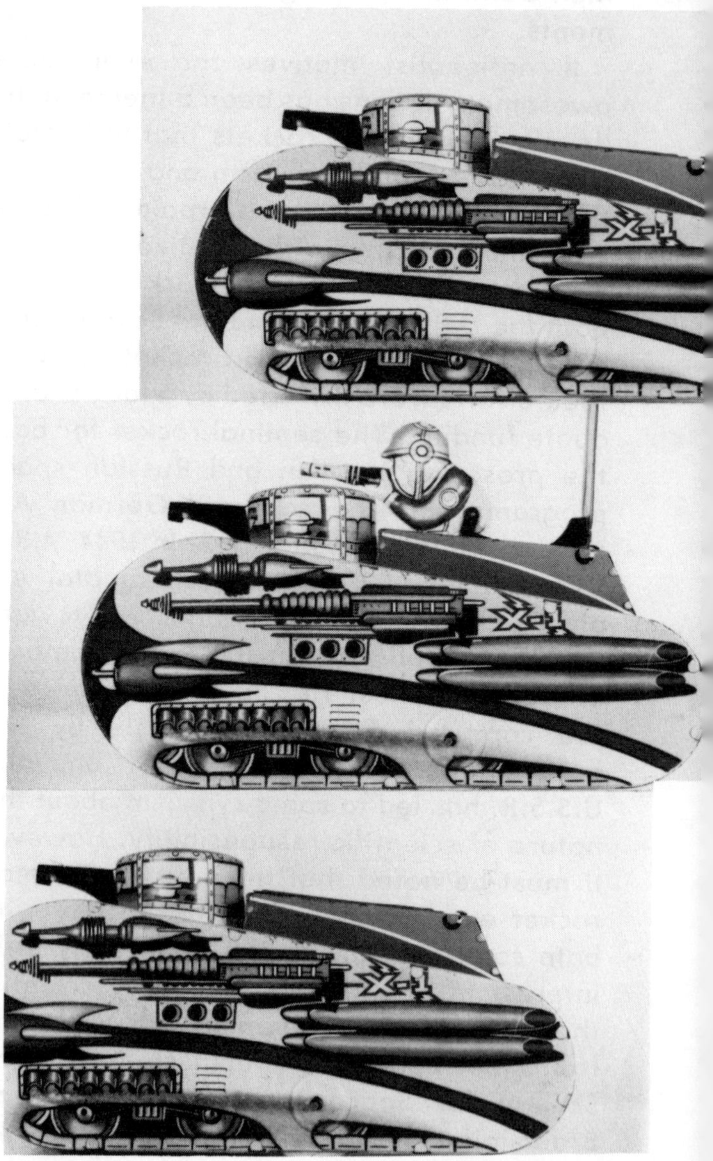

X-1 SPACE TANK. This tin tank by Marx was made for years with a number of different lithos. This one is of REX MARS, a personnage whose other exploits are unknown to us.

COUNTDOWN. The widely publicized ritual of the N.A.S.A. countdown prior to rocket lift-off inspired a number of miniature toys that launched projectiles. Many have TV screens and reference to clocks and other features of the control panels.

JUPITER SPACE STATION. Like the COUNTDOWN this Japanese battery operated toy mixes the realism of the 1960's space program with the hokey charm of the 1950's juvenile TV space shows. Our example, like most, is missing the rockets. Perhaps it worked better than we thought and they are still in space.

COUNTDOWN. This large plastic set by Ideal transports any of the 3 rockets along a track to the launch site. There the countdown begins resulting in a blast off. The toy is illustrated below the box. A detail of the countdown clock is shown, below right.

SKY SWEEPER. Many American rocket toys have a military flavor. This all plastic ROCKET LAUNCHER by Ideal projects 6 different images and fires suction cup rockets at them.

CAPE CANAVERAL MISSILE BASE. Marx made a number of large playsets with a space theme in the 1950's and 1960's. Others were called OPERATION MOON BASE and CAPTAIN SPACE SOLAR ACADEMY. At least three variations of the CAPE CANAVERAL set exist all filled with countless plastic men, equipment and tin buildings and fences. They capture a lot of the hot-dog, A-OK flavor of American public enterprises of the post-World War II, pre-Vietnam era.

ROCKET LAUNCHER. Real 1930's flavor graphics in a 1940's game. The plastic six gun reminds us that spacemen and cowboys are members of the same brotherhood — the upholders of truth and justice and defenders of women and children from the dusty plains of Texas to the coldest moons of Saturn.

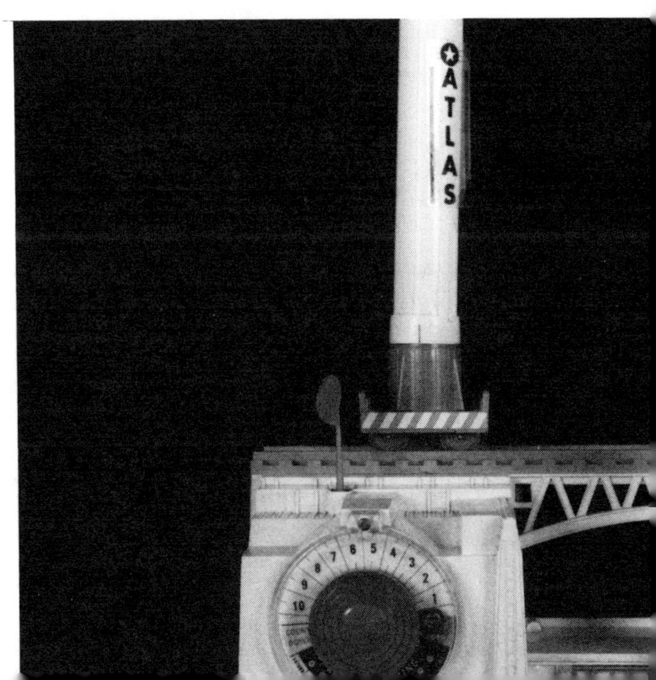

FACE OF THE MACHINE

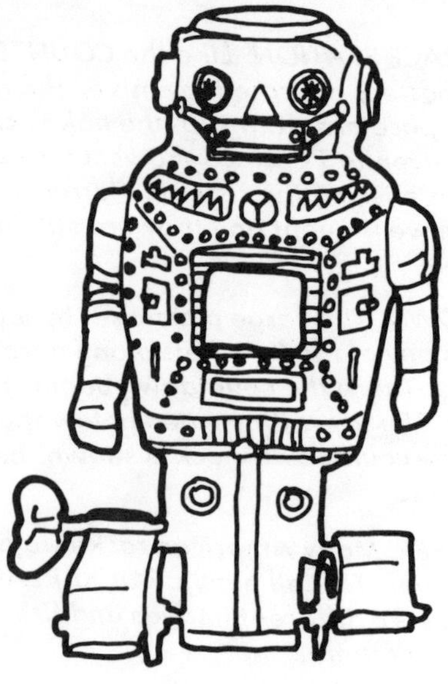

KEY WIND ROBOT. Numerous versions exist of these small tin robots with a sparking chest. None are menacing but this one is positively friendly.

STAR WARS RADIO CONTROLLED R2-D2. New in the 1978 Kenner catalogue and continued with minor changes for 1979 and 1980, this 8" tall all plastic radio controlled toy is an accurate replica of the lovable little robot in George Lucas' STAR WARS. With its small size, "beeping" speech and cute actions R2-D2 is more of a pet than a person. It, or "he" as the Kenner ad copy refers to the robot, is automation at its least threatening.

There are a few clown-like toy robots - the Mego Man, Lucky, the Driving Robot among them. At the other end of the scale there are not many monster-like robots - Dino, the leering, flesh eating dinosaur in a space suit and the bizarre Thunder Robot by Asakusa, for instance. Most of the toy machines with faces are expressionless, however. They give no cause for either joy or fear. The basic Japanese toy robot is stiff and formal but reliable and reassuring in a curious sense. Many toy astronauts by contrast carry guns, but fewer robots do. The classic Japanese robot toys are by and large not hostile. They are machines without malice.

In Western society we alternately love and hate the machine and its parents - science and technology. Improvements in technology can provide better and cheaper food, shelter and living conditions. Technology has progressively delivered more and more lethal weapons as well. Faster automobiles cause more deaths on the highways. Some associate the scientifically run factory with poisoned air and water. Not only are atomic weapons opposed but considerable numbers of people want nuclear power plants abolished. When Detroit announces automation of a factory unions are quick to protest. These anti-machine attitudes go back to the early industrial era when London seamstresses smashed sewing machines fearing they would be replaced by them. Our literary history is full of Frankensteins, run-away robots and even killer computers. To be sure there is the gentle giant, Robby, in FORBIDDEN PLANET and R2D2 and C3PO in STAR WARS who between them combine the all around usefulness of the ultimate Swiss army knife with the social graces of an English butler. More often our science fiction creates alarming and destructive robots just as our society sometimes expresses fear and loathing for machines.

Evidently the Japanese culture sees machines in a more favorable light and perhaps that cultural bias is reflected in the enthusiasm and cleverness with which they satisfied the world market for space toys for 20 years. At any rate the center of toy robot production has moved from Japan to China and Taiwan. The Japanese now lead the world in the production of programmable industrial robots. According to the Dec. 13, 1981 NEW YORK TIMES Japan has 14,000, the U.S. 4,100, West Germany 2,300, France 1,000, Sweden 600, and Britain, home of the Industrial Revolution, has 500. There are now operating in Japan several factories where robots make other robots. At the $38 million Fijitsu Fanuc Ltd Tokyo plant robots produce 100 more robots a month. Far from fearing replacement by the robots Japanese workers view the new devices as welcome helpers performing routine, boring or dangerous jobs. At the Yamazaki plant near Nagoya a transport robot that lifts 16 ton loads has a Popeye painted on him as though to equate the machine with the good hearted but very strong cartoon character. Gerard Perron notes in Boogaerts' book ROBOT that these sophisticated industrial robots even look like enlarged versions of their toy robots of 20 years ago, vaguely human but not competitively so. They are the good robots, the reliable, smart machines, not the distressing look-alike androids - smooth skinned, fast talking immigrants of our imagination who have an eye on our jobs, our wives and God knows what else.

All of us are fascinated by machines that move like men or men that move like machines - i.e. Charlie Chaplin in MODERN TIMES and the rock group Devo, today. The philosophical ramifications of this are enor-

ATTACKING MARTIAN AND RADAR ROBOT. Some Japanese robots have human heads or faces. The comingling of human and mechanical features in toys can be seen as a reflection of advanced industrial society where automation blurs the boundaries of what is human and what is machine. Depending on our focus we may be confident at the prospects of sharing our work with robots or fearful that we may be supplanted.

mous. Since robots are the premiere space toy category we should become aware of the history of robots (or automata as they were long called) and the dangers and delights they have variously seemed to present. Robot use in industry to replace human workers promises to be one of the major issues of the 1980's. Hardly an edition of TIME or NEWSWEEK goes by without a robot article. In particular the American press is excited (and repelled, too) by the before mentioned Japanese robot factories. The industrial robot industry began in the U.S. but American firms not only employ (if employ is the proper word for a worker that receives no paycheck, never goes on strike and takes no coffee breaks) far fewer robots and they will fall much farther behind in the future. In Japan there are 140 robot manufacturers compared with 20 in the United States. A securities analyst predicts that by the end of the decade the output of robots in Japan will be "4 times greater in units and value than the most optimistic forecast for the United States." By the end of the century Japanese robot population may be one million.

To be sure concern for the unemployment that automation temporarily causes is one explanation for the backwardness of the West in this area of technology. Even greater than the fear of death may be the fear of being rendered obsolete or useless, "redundant" as the English would say, while still in one's prime. We all must die but unemployment is a particular kind of hell that because some avoid it, the losers feel damned. Being replaced by a machine cannot but add to the insult. Our society has had a long running battle between its religious authorities and those with a mechanistic view of the universe. The position of Christianity is that only man has a soul and all other creatures are therefore inferior. Imagine the status of a machine if even an animal is without the blessing of the Creator. The knotty topic of machines that can perform human functions was discussed in a book published by the French philosopher Rene Descartes in 1637. Complex spring driven figures capable of playing musical instruments and walking had been built and widely displayed as early as the middle 1500's. Such automata convinced the public that the day would come when man could build fully functioning replicas of humans or animals. There are apocryphal stories from this era about androids that would go out and do their makers shopping in the town square. Decartes himself is rumored to have built a female android he named Francine who travelled with him discreetly in a trunk until a ship's captain pitched her overboard in a storm. Prejudice against robots is not new in our history. Descartes incidentally concluded in his treatise on robots that man will one day perfect "soulless machinery" capable of independent motion but that God intended man alone to possess a soul.

The precise origins of the word robot is an illustration of another negative European discussion of the place of automata in society. The medium is a play by a Czech, Karel Kapek, called R.U.R. (ROSSUM'S UNIVERSAL ROBOTS). Written in 1921, 284 years after Descartes' pronunciations on the spirituality of the robot, Kapek passes judgment on the economic value of automata. He concludes as well that the robots have no soul, do nothing but work and will ultimately take over, destroying humanity. It is a parable of Kapek's negative feeling about capitalism to be sure but his fears about automation are genuine and prove how deeply rooted and widespread misgivings about the machine are in the West.

The words of the remarkable Osamu

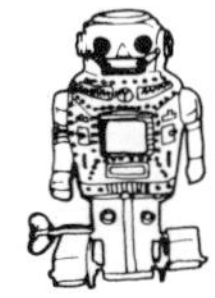

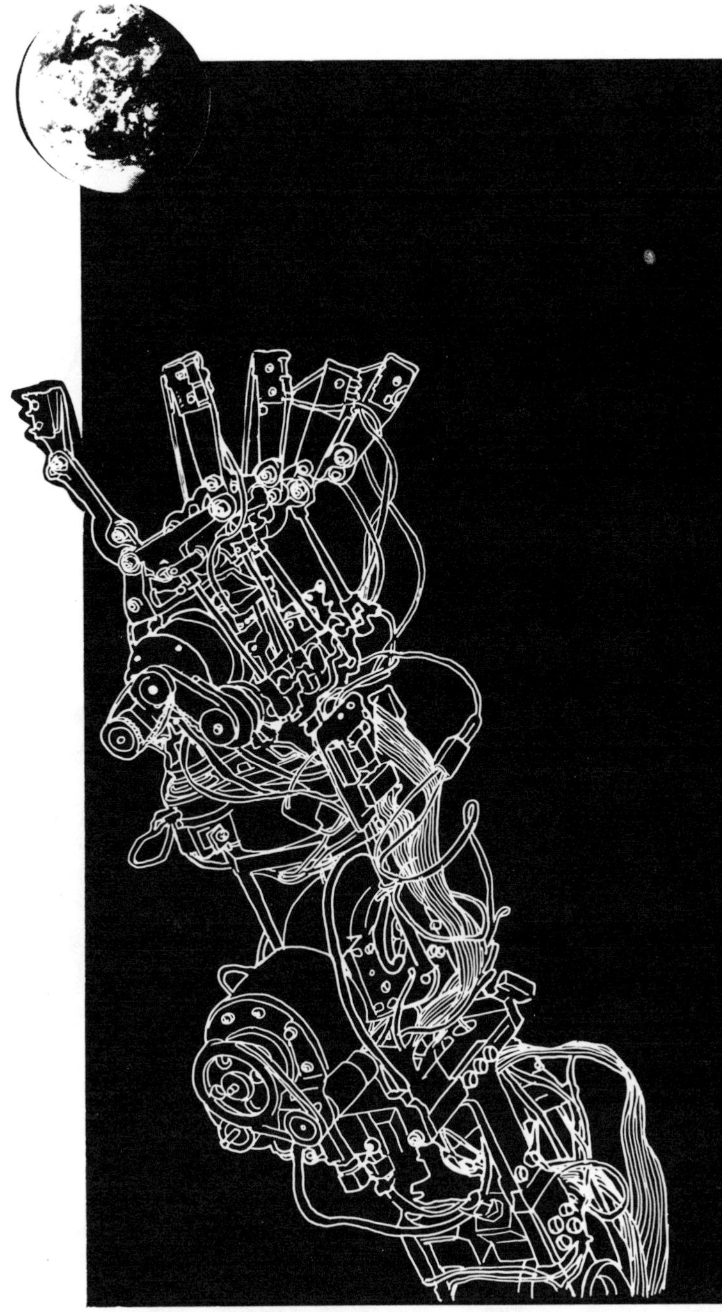

Tezuka called the "king of Japanese comics" points out the difference in the two cultures, explaining at once why the Japanese space toys are classics and why the Japanese are also pace setters in automation. In the NEW YORK TIMES Sunday Magazine article called "Japan's Love Affair with the Robot" Mr. Tezuka explains: "Unlike Christian Occidentals, Japanese don't make a distinction between man, the superior creature, and the world about him. Everything is fused together, and we accept robots easily along with the wide world about us, the insects, the rocks - it's all one. We have none of the doubting attitude toward robots, as pseudohumans, that you find in the West. So here you find no resistance, simply quiet acceptance."

It is reported that Shinto priests in silk robes are sometimes called upon to wave their wands over robots in a purification ceremony attended by the entire staff and management before they are put to work in Japanese factories.

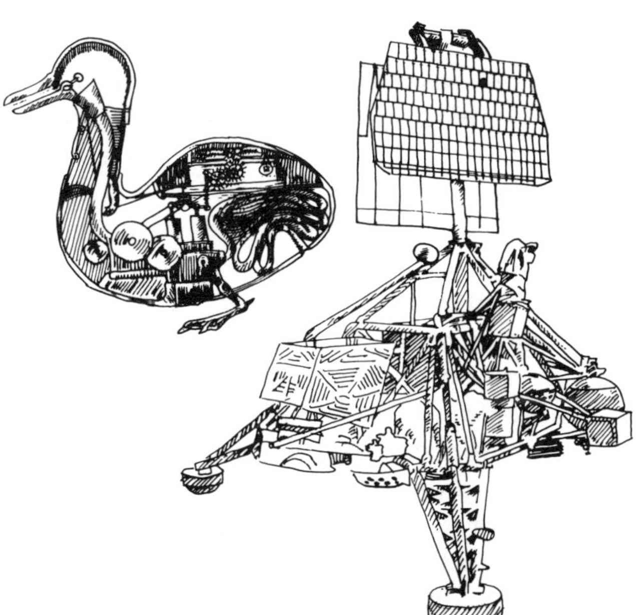

(ABOVE) ARM OF AN INDUSTRIAL ROBOT. Japan leads all the countries in the world in the production of industrial robots. The United States was responsible for their invention and perfection but the Japanese have been the most aggressive in employing them in factories. Could the charm of Japanese toy robots reflect their culture's easy acceptance of mechanical beings? A study of popular culture may provide some insights into the attitudes of the different peoples of the world and why advanced technology is accepted by some and rejected by others.

(LEFT) DUCK AUTOMATON, 1738, AND MOON LANDER, 1969. Machines that function automatically have been a preoccupation of technological man. Entertainments such as the automatons preceeded useful application of robotics by centuries.

SCIENCE FICTION AND POP CULTURE

FLYING SAUCER. This 1950's science fiction toy shows a generalized, almost astrological, treatment of the heavens. Collectors do not value such non specific designs highly. The best toys have the tart flavor of that period's science, fiction or pop culture. A study of these associations will reveal tantalizing parallels and give new enjoyment to one's collection.

MICKEY MOUSE AS CYBORG. Popular culture endlessly revises the meaning of all its characters. Nothing is safe from updating. Mickey Mouse here is a cyborg — a living creature with mechanical replacements of vital organs.

It is increasingly recognized that there are within the body of writings known as science fiction important works of literature. Not everyone agrees (nor does everyone agree as to what science fiction is or literature for that matter) but it's an idea of sufficient currency to generate several books a year arguing the point and countless articles and theses. Naturally this little toy book has nothing to add to this. We accept Isaac Asimov's concept that science fiction began in the early 19th century as a "literary response to a new curiosity" about the effects of change on humans and society caused by scientific and technological progress. Not all science fiction takes place in space but most of it chooses for a locale some place other than the earth in its present state. Almost always the science fiction work takes place in the past, the future or far away. The literary device of spinning a tale in a distant place was recognized by the ancient Greeks. The ILLIAD and the ODYSSEY had horrible monsters and wondrous nymphs no Athenian had ever seen but of course Ulysses' adventures were far, far from home. (The ILLIAD and the ODYSSEY are not science fiction, Asimov concludes because they operated in a prescientific world controlled by supernatural forces. Frankenstein was animated by electricity, not magic).

To say the least, space has been a very popular stage for the science fiction writer. The style and content of space literature ranges from profound to puerile. But some of the less thoughtful has been the source of mass entertainment. Aside from the specialized appreciator of pop culture as "kitsch" or "camp" we must acknowledge the effect that science fiction, good, bad or indifferent, has had on our perceptions. Our history has been affected by it all. Space toys are an example but they are only a spin off of this large body of science fiction. Let us leave literary judgments to academicians and concede that science fiction - all modes of it - has been an ingredient of popular culture for some time.

What have been the messages of these novels, stories, comics and movies and how have people received them? What has been the origins of their ideas in science and fact, if any. And finally what about trickle down theories and spin offs where a concept moves from one medium to another. This last evolution will be of special interest to collectors because it frequently explains how the toy came into being.

There is sometimes feedback. Science fiction and the pop culture versions may actually influence real events. Such an example might be the saga of atomic energy. In a high minded way the Atomic Energy Commission has promoted the use of nuclear power but the hearts and minds of Americans were captured by grade C horror movies of the 50's and the 60's in which giant cockroaches and 40 foot tall lizards came glowing out of the deserts after a brush with radioactivity. No doubt there are serious critics of nuclear power who have founded their charges on scientific facts but it's hard to dismiss the enormous anti-nuclear propaganda force of those films. Now the kids who watched the glow-in-the-dark reptilian mutants eat an entire 1954 Corvette and its teenaged occupants are voters and protestors to be reckoned with. It's pop science fiction creating history.

Childhood playthings admittedly don't lend themselves to horror themes but another 1950's pop science fiction cliche, flying saucers, inspired numerous toys. About the time journalists began reporting that rockets and rocket planes were about to actually skip out of the earth's gravitational field and fly in space the public began

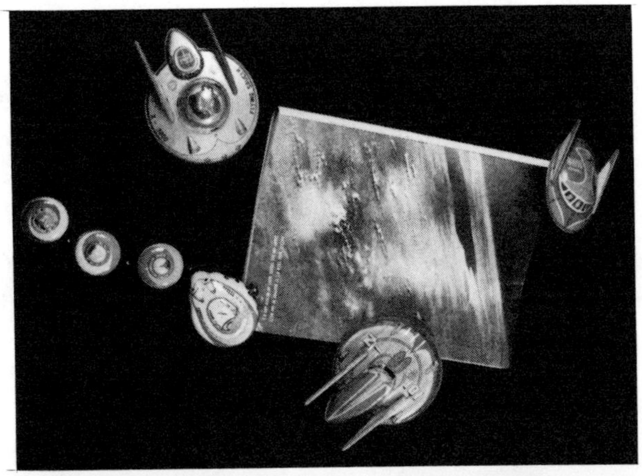

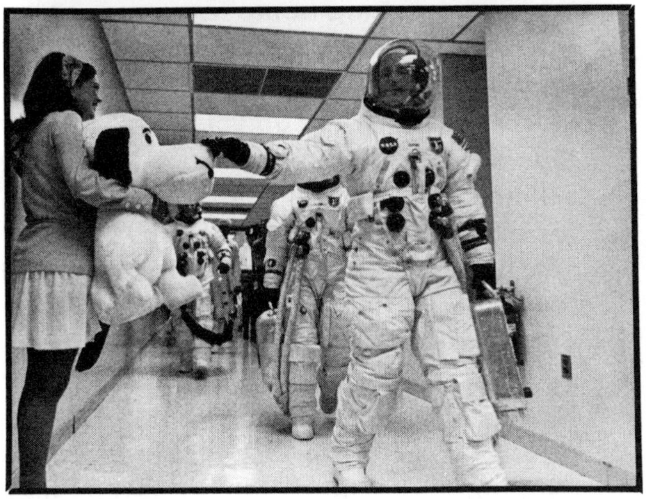

TOY FLYING SAUCERS FROM THE 1950'S. As rockets began to bring back actual photographs of the earth and heavens from space some of the most simplistic science fiction began to look ridiculous even to children. CAPTAIN VIDEO and B- movies with bug-eyed monsters on the moon were casualties of the growing awareness of the scientific realities of space. Flying saucers however have proved to be a resilient pre-space travel bit of popular mythos. No astronaut has ever seen a saucer, no actual scientific evidence of saucers exists yet there is not a toy store on earth that doesn't stock some kind of flying saucer. After Stephen Spielberg lands his very 1950's looking saucer in E.T. THE EXTRATERRESTRIAL we can expect a new wave of saucer sightings and new squadrons of saucers will begin flying in from the Far East toy factories.

APOLLO 10 DEPARTURE. The Apollo 10 mission preceeded the actual landing on the moon. During the orbital operations the Lunar Module was called "Snoopy" and the Command Module was called "Charlie Brown." Even at N.A.S.A. popular culture mixes with science.

looking at the skies with renewed interest. Ancient man spent much time gazing upwards for signs of weather changes and possibly to catch a glimpse of a low flying celestial being. In the 20th century our weather report comes to us from a radio and scarcely anyone believes Zeus will part the clouds and look down upon us. When people looked into the early 1950's skies some of them saw strange things. Waitresses going home late, cops patroling the edge of town, even airline pilots saw unexplained phenomena in the sky. These unidentified flying objects (UFO's) may or may not have been real but a large body of science fiction of all levels was ready to offer explanation. Back as far as 1919 an eccentric newspaper reporter, Charles Hay Fort, began clipping curious stories from the pages of history and publishing them. In volumes such as *The Book of the Damned, New Lands,* and *Lo,* he wrote about these mysteries offering no explanation of his own but subtly implying that reality may be a lot less real than we think.

Fort died in 1932 but he opened the way for others to begin to mix science fiction up with the news. First Edmond Hamilton corresponded with Fort after reading his books and in a note to his 1930 story "The Space Visitor," Hamilton concluded Fort had offered evidence that earth had been visited for over 150 years by "extra-terrestrials." These claims sold copy especially in the pulps and a cult grew around the idea we were and are periodically visited by aliens from outer space. (In recent years Erich von Daniken, in books like CHARIOTS OF THE GODS? and RETURN TO THE STARS has continued the tradition of cashing in on the public's inability to always tell the difference between science fiction and reality.

This being so, how natural it would be for aliens to step up their surveillance of earthlings right as we were about to come out into space ourselves. Public concern over atomic power was supporting a healthy horror industry already and the visiting alien theme also found ready mass acceptance. The two ideas combined in the 1951 film THE DAY THE EARTH STOOD STILL by Robert Wise. In it we are warned by a very gentle alien to cool it with atomic violence or else. Gort, his 12 foot tall robot that vaporizes whatever the boss doesn't like, leaves no doubt that we better change our ways. The next year another kind of alien lands in a flying saucer but he, or "it," brings no moralistic plea for peace. In THE THING FROM ANOTHER WORLD credited to Christian Nybe but directed by Howard Hawks, the alien is a plant who drinks blood. The thing regards humanity as we regard an orange grove - just something to use for juice. Generally speaking you wouldn't want whatever got off a flying saucer in a 50's film for a next door neighbor.

Toy makers got the word. Not only were the flicks full of saucers, the newspapers had a UFO sighting every other day. Lindberg Products came out with a 10 piece plastic Flying Saucer kit in 1952 with a little green creature to fly it and an unreadable foreign decal. We haven't determined when exactly the Japanese began producing battery operated toy saucers. They must have found a receptive market for they have produced hundreds of different variations. With new lithographs and cosmetic changes alien flying saucers could become in time satellites and space ships very definitely piloted by our astronauts. In the cockpit of some are baby faced space explorers. Others are driven by serious sorts but even the older saucers with little green men types are not at all like "The Thing." Strangely the atomic symbol appears often on the Japanese saucers as well as other space toys of theirs. It seems remarkable after the traumas of Hiroshima and Nagasaki. While there was little optimism in the science fiction movies of the 1950's the toys, undoubtedly meant to cash in on the subject matter, are often uncommonly jaunty and cheerful. Perhaps it's the nature of toys to aim at an audience not ready for the more disturbing things in life but in part it may be due to the enthusiasm the Japanese culture as a whole feels for advanced technology. More than anyone else at this time Japan seems positive about the benefits of science and the spirit of the toys don't conceal it.

BOTTOM OF BOX OF INTER-PLANET-TOY ROCKETANK PATROL. The picture on the bottom of this 1950 MacRey toy shows a typical American science fiction theme — space creatures gleefully destroying cities. On the whole America's popular culture science fiction depicts more wantonly aggressive aliens and robots and science fiction monsters than Japanese popular culture. Even Godzilla had a heart. The Marx, made in Japan, MECHANICAL MOON CREATURE looks more like a cultivated tourist than the town stomping monsters on the American toy box.

STRADER AND MEKANDA ROBOT. A STAR WARS figure encounters a contemporary Japanese diecast toy. The contribution of the many levels of science fiction and space events to our subconscious is not to be underestimated. Taken collectively these influences may shape our culture more than is recognized. Space toys are not a trivial collecting interest for they can conjure up this rich phantasmagoria.

THE PULPS

PULPS. The inexpensive escapist male reading material called "pulps" flourished in the first third of the century. Along with westerns and horror stories space tales were a popular pulp genre.

SPARKY ROBOT. The excellent cover art work of the pulps has doubtlessly had an effect on toy design. It's easy to see a pulp cover influence on this 1960 · Japanese robot and its box.

Space toys inspired by the more traditional literary forms of science fiction are almost non-existent. Novels that handle the sober themes of man's place in a scientifically constructed universe are by their nature somber. In such adult works by celebrated writers such as Aldous Huxley, H.G. Wells, Franz Kafka and in our day, Kurt Vonnegut, modern problems of alienation, depersonalization, and existential anxiety follow man wherever he goes into time, space or introspection. The universe opened up by the literary lions of science fiction has influenced popular science fiction but that is a matter beyond the scope of this toy book. It is germane to consider the pulp era of science fiction.

"Pulps" (getting their name from the cheap wood pulp paper they were printed on) were usually 7 by 10 inches with a ½ inch spine. Out of them came the character of Buck Rogers. The colorful covers of the pulps with artists' renderings of robots, spaceships and ray guns have no doubt been a major influence on toy designs. Most importantly the early family pulps included a scattering of science fiction stories which introduced space and time travel to a wider audience than most of its previous incorporations in novels (although a few novelists like Jules Verne and Wells enjoyed mass popularity). When specialized pulps came along they, in a sense, invented the category itself, promoting criticism and definition.

The pulp version of space and robots and rockets is the one that Hollywood and the comics took as a starting point. Even movies made from serious novels more often than not take on a pulp flavor. Among the various genres of pulps - western, sports, detective, adventure, supernatural, girlie and science fiction - there is much common ground. Scantily clad white girls in peril (menaced by Mongol bandits in adventure pulps; threatened by bug-eyed aliens in science fiction), exotic locales (jungles, distant galaxies - anywhere but Cleveland in the 30's), violence (knives, guns, death rays) and ever the strong, masterful hero (Buck, Tarzan, Nick Carter). Perhaps pulps tell us more about their audience than about their authors. Science fiction pulps were predominantly male-oriented escapist reading material. Who cannot notice the emergence of the world-saving superhero in the Depression - when many male readers couldn't even feed their families, much less set the world right? In the early years of the century pulps' heroes, by contrast, were strong but they lived apolitical lives.

Before ARGOSY, the first pulp, was published in 1896 by Frank Munsey there were dime novels. Published weekly, usually costing a nickle, they aimed rough and ready stories at boys and young men. Not a few of the stories had a science fiction flavor - usually featuring spunky lads, advanced technology and adventure in far away places. BEADLE'S AMERICAN NOVELS, such a paper, published in 1868 "The Steam Man of the Prairie" by Edward S. Ellis, concerning a steam driven robot who goes west and fights Indians. Frequently the protagonist was a young inventor himself and often he used his futurist device to fight Apaches or explore the North Pole. Topical and escapist at once ARGOSY was originally such a boys' magazine. As an adult pulp magazine it remained topical and escapist but dealt with older characters and less juvenile situations. Although blood flowed like rivers from the pages of the early pulps, male/female relationships were by our present standards immature. Sex didn't surface until the specialized pulps of the 1920's and 30's and even then it was tame by today's standards.

By then competition from the movies, radio and picture magazines had taken pulps' general audience. The naive attitudes of the early pulps were challenged by a spectacle of social unrest, labor troubles and the horrors of World War I - to say nothing of the paper shortages caused by the war. ARGOSY merged with ALL-STORY and CAVALIER to become ARGOSY-ALL-STORY WEEKLY. In 1920 it cut its pages from 224 to 144. The time had come for the genre pulps - periodicals which appealed to a narrow but loyal market. Topics among these specialized pulps were as mentioned - sports, westerns, "spicey" stories, adventure, supernatural and science fiction.

On April 5, 1926, Hugo Gernsback published the first all science fiction magazine. It was 8 X 11 (larger than the standard pulp) and it was called AMAZING STORIES. It consisted of reprints of Wells, Verne and Poe for the first two issues. As manuscripts were submitted AMAZING published new material. In 1930 ASTOUNDING STORIES OF SUPER SCIENCE challenged Gernsback's monopoly of magazine science fiction. Others followed although by now the Depression was making 25 cents a lot more money than it had been in the 20's. Gernsback had sold AMAZING in a 1929 bankruptcy suit but had a similar new pulp, SCIENCE WONDER STORIES, ready a month after his last issue of AMAZING. In response to the hard times he trimmed the size down to popular pulp dimensions (AMAZING had done so too in 1933) and in 1935 he cut the price and tried to sell it by subscription through the mail to avoid splitting the profits with newsstands. He failed and sold WONDER to Standard Magazines and for a time left science fiction publishing.

Gernsback came along at a time in which

popular science fiction was of two categories. One was the very romantic, featuring strange, distant worlds, larger than life heroes and villians and beautiful innocent maidens forever in peril. Edgar Rice Burroughs' 75 books all have this post Gothic novel flavor. Burroughs' Mars is not a part of the solar system - it exists only in his imagination and in the imagination of a sympathetic reader.

Gernsback's preference was for works of fiction in the second category, stories that were excessively gadget-y. He had first published a radio magazine, MODERN ELECTRONICS, and in his magazine's "gee whiz" technology counted for more than the appeal to the primitive that Burroughs' fantasy embodied. Magazine science fiction was to become considerably more sophisticated in the late 1930's as writers and audiences contemplated a world not made perfect by either tough talking individualists or "gee whiz" technology - the two implicit heroes of so much of the earlier stuff.

A synthesis of Gernsback's "popular mechanics" kind of science fiction and the more emotionally fulfilling science fiction fantasy of Burroughs came about in the late 1930's in ASTOUNDING SCIENCE FICTION magazine under the editorship of John W. Campbell, Jr. Hardly a book on the history of the genre fails to devote a chapter to this man and the influence he had on the form. Campbell, himself a capable writer (under the pseudonym of Don A. Stuart), brought together and consistently published many of the writers who are today considered "names" in the field. These included Isaac Asimov, L. Sprague de Camp, Robert A. Heinlein, A.E. Van Vogt, Lester del Rey, and Theodore Sturgeon. Although considerable individual differences exist in their work they all strive for scientific technical plausability without sacrificing the human interest angle of the story. From the late 1930's to the disruptions caused by the Second World War a renaissance occurred in small science fiction magazines. Although pulp magazine science fiction opportunities would decline many of the writers who started in the "golden years" of science fiction pulps had sufficient stature to become publishable in slicks, hard cover books, some even went on to write for TV and movies. By 1944 all but 7 of the 18 pulps that had thrived in this boomlet of science fiction were gone. Comic books were in flower then and they siphoned off not only the kids but probably a surprising number of literates who might have once been readers of the short stories in pulps. The size of the pulps shrank again to 5½ X 7¾. Competition from paperback books in the 50's further cut into the mature magazine market and television would soon offer juveniles pop science fiction entertainments even easier to digest than comics. The print media in general has progressively surrendered ground to picture books, movies, TV and radio. Important science fiction continues to be written and published but little of it directly affects popular culture today. It must go through another form to reach a mass audience, movies or TV usually. As we are only interested in the places where science fiction influences toys or the market for toys we need not vigorously pursue books and magazines of the last 30 years.

Gernsback tried to re-enter the field in 1953 with a color slick SCIENCE FICTION PLUS. It lasted 7 issues. Edgar Rice Burroughs died three years before in California at the age of 74. He was reading a comic book in bed.

There is pro-technology science fiction and anti-technology science fiction.

Science fiction that openly admires technology would seem more likely to produce the most toys. Buck Rogers, for instance, was a character who wouldn't be caught dead without an Atomic Disintegrator Pistol at his belt while Flash Gordon, derived from a more Burroughsian anti-technology fantasy form, could lash out effectively at his enemies with medieval weapons. Although many other factors are to be taken into account Buck's toys are much more abundant than Flash's. More recently, STAR TREK produced a lot of licensed products but few actual toys compared to STAR WARS which may become the biggest line of character related toys in history. STAR TREK while not anti-technology in a Burroughsian sense is large concerned with problems usually solvable by psychology. STAR WARS cannot hide its makers enthusiasm for technology. The designers and special effects people within the Lucas organization lavish attention on every piece of hardware with the result that the toys derived from the movie exhibit a variety and abundance of form highly pleasing to the children — who are all enamored at this stage of their lives with the benefits and power of gadgets.

SPACE COMICS

LITTLE NEMO. From 1905 to 1911 R. Winsor McCay drew a fantasy comic strip that often contained reference to space travel. Through the comics the public became aware of science fiction situations even before the pulp hero Buck Rogers came to newsprint.

The essential space hero is Buck Rogers. In an August 1928 AMAZING STORIES pulp magazine there was a story entitled "Armageddon 2419 A.D." by Philip Nowlan. It was about Anthony Rogers, a 29 year old U.S. Air Force Lieutenant who was put into a coma by radioactive gasses in a mine shaft and came to in the 25th century. Finding the world overrun by "Mongols" and ruled by merciless dictators Rogers took immediate action to set things right, helped of course by the spunky and lovely Wilma Deering. John F. Dille, head of the National Newspaper Service was looking for a new action strip. He suggested they change Anthony's name to Buck and he teamed Nowlan up with a W.W.I Army pilot who could draw named Dick Calkins to create the space adventure strip which first appeared on January 7, 1929. Its public acceptance spawned FLASH GORDON in 1934 - the year Daisy Manufacturing Co. introduced its first BUCK ROGERS Space Pistol, the X2-31 - a heavy gauge blued steel wonder with nickle plated trim that goes "zap" when fired. Nearly three million guns were sold starting the avalanche of BUCK ROGERS character merchandise that remained unequalled in space toy history until STAR WARS over 40 years later.

Tying in toys and other merchandise to comic strip personnages was not new. AT THE CIRCUS IN HOGAN"S ALLEY by R.F. Outcault first appeared in the Sunday NEW YORK WORLD on May 5, 1895. In it a jug-eared lad of 6 appears in a dirty blue nightshirt. By January, 1896 the lad had become bald, his smudged nightshirt was yellow and the strip was called HOGAN'S ALLEY. The "yellow kid" as the public called him was the original smart alec talking back to the public in a highly irreverent tone. When Outcault went to William Randolph Hearst's NEW YORK JOURNAL the title was changed to the YELLOW KID and the public was offered dozens of kinds of trivia with the picture and name of America's first comic character. These include puzzles, games, buttons, books and a number of toys - a cast iron cap bomb and a cast iron goat cart driven by the Kid among them. The strip was influential but short-lived due to middle class offense at the frankly depicted slum environment and the salty characters. Litigation between the JOURNAL and the WORLD over the rights to it gave us the phrase "yellow journalism." Outcault went on to create the BUSTER BROWN Sunday strip in 1902 and became wealthy by 1910 as much from the marketing of products carrying the Buster Brown logos as the comic.

As the early comics established the business of licensing characters to manufacturers of toys, clothing and other merchandise, they also ventured into space. LITTLE NEMO IN SLUMBERLAND created by R. Winsor McKay in the Sunday NEW YORK HERALD from 1905 to 1911 often had dream sequences in which the hero, Little Nemo, encounters strange beings while blasting through space on a rocket. More a personal fantasy than a derivation of science fiction this presurrealist masterpiece reminds us that the masses had seen space as a background very early on. It also points out the link between our subconscious and events in space such as weightlessness, contradictory perspectives, ambiguity and alien beings - all stock and trade devices of future science fiction comics.

BUCK ROGERS

Conceived of in a pulp science fiction magazine, nurtured in a newspaper comic strip, Buck Rogers grew to occupy a prominent position in films, radio and television. The periodic nature of comics and their graphic form keep a figure in the public eye

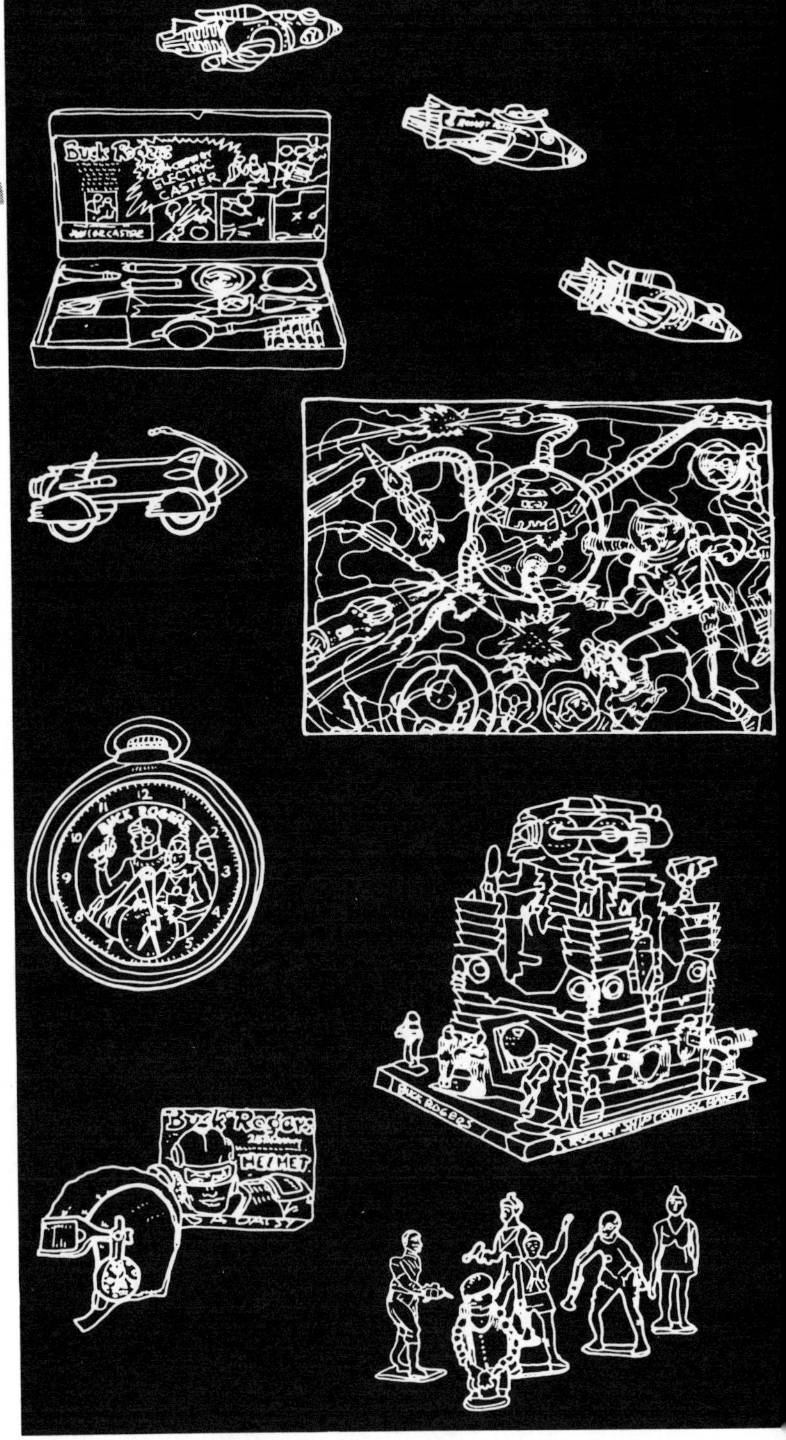

BUCK ROGERS ITEMS. (Top to bottom) Three versions of Buck's 1930's ROCKETSHIPS by Louis Marx. ELECTRIC CASTER to make lead figures. WAGON shaped like rocketship. PUZZLE by Milton Bradley, 1952. POCKET WATCH, 1935. DR. HUER'S LABORATORY by Warren Paper Products, 1937. HELMET by Daisy, 1934. METAL JOINTED FIGURES by Britains, radio premiums from Cream of Wheat in 1930's.

and lend themselves to merchandise endorsement. BUCK ROGERS is one of the great success stories of all times in this regard. In his various revivals in the last 54 years hundreds of items have had a Buck Rogers tie-in - from wrist watches to sneakers. Let's follow the remarkable history of BUCK ROGERS, Space Adventurer (and character merchandiser) Extraordinaire.

As we mentioned Buck first appeared in a 1928 short story by Philip Nowlan in AMAZING STORIES. National Newspaper Service got W.W. I pilot, Dick Calkins, to draw it and on January 7, 1929 it first appeared in the funnies - ironically on the same day as the first TARZAN strip. Edgar Rice Burroughs' jungle man is an ode to the virtues of the primitive while Buck extols the technology of the future. The first two adventure strips both have been tremendously resilient concepts. Tarzan however has spawned few toys. A man in a loin cloth who often rejected even a knife when killing a crocodile or a lion doesn't inspire toy makers like the 25th century hi-tech adventurer. You don't swing on vines between stars and it's hard to duke it out with bad guys who rocket around at the speed of light in spaceships. Gadgets are the answer and kindly Dr. Huer could crank them out of R. & D. and into production almost as fast as toy makers could hit the market with goodies labeled "Buck Rogers". Many of the 1930's items may have been made in fairly small numbers or been a bit expensive for they don't surface much today. Even several of the give-aways are scarce and valuable on today's market. Hard playing alone may account for the rarity of some of the Buck Rogers items.

The daily BUCK ROGERS IN THE 25TH CENTURY strip originally drawn by Calkins has been worked on by other artists - Murray Anderson in 1947, Leon Dworking in 1949, Rick Yager in 1951 and Anderson again in 1958. Finally George Tuska worked on it until its finish in 1967. Besides Nowlan other writers include the illustrator Calkins (1940-1947), Bob Barton (1947-1951), another artist Rick Yager (1951-1958) and others including Fritz Leiber.

On March 30, 1930 a Sunday BUCK ROGERS began which ran until 1965. Never drawn by Calkins (but signed by him) it at first didn't even contain Buck. His girlfriend, Wilma Deering's kid brother, Bud Deering and his girlfriend, Princess Alura from Mars starred in this. Buck came back years later and took over. Among the artists on the Sunday strip were Russell Keaton (1930-1933), Rick Yager (1933-1950), Murphy Anderson (1958-1959) and George Tuska (1959-1965). The writing was by Phil Nowlan (1930-1940) and Rick Yager (1940-1958).

Buck has been published in comic book form on several occasions. The first is a 1933, scarce 6 X 8 Kellogg's Corn Flakes give-away, BUCK ROGERS IN THE 25TH CENTURY, by Nowlan and Calkins. In the first years of comic books the strips were reprinted from the newspapers. The first monthly newsstand comic book was called FAMOUS FUNNIES and in the third issue (1934) the strip reprint of BUCK ROGERS began. In 1937-38 Vicks Comics reprinted the early Buck strips again. Buck got his own cover in BUCK ROGERS COMICS of 1940, 1951 and 1964. The fantasy artist, Frank Frazetta, did some of the drawings in these later ones.

Veteran movie serial director, Ford Beebe, cast Buster Crabbe in the role of Buck in a 1939 serial version. FLASH GORDON, created as a strip 5 years after the appearance of Buck in the papers, beat Buck to the silver screen by 3 years. Moreover Buster Crabbe also starrted as Flash. That must have caused some mini-identity crises among the junior space set back then. Actually all the movie serials were so recycled (and even retitled) that kids up to today have had a chance to see Buster as Buck or Flash zapping the outer space evil doers on the tube.

In the flood of kiddie TV space programming of the 1950's Buck Rogers appeared as a short lived series by ABC. First telecast on April 15, 1950, it couldn't take the pressure from CAPTAIN VIDEO and was last shown on January 30, 1951. Buck was played first by Ken Dibbs, then Robert Pastene replaced him. Wilma Deering was played by Lou Prentis, Dr. Huer by Harry Sothern and Black Barney by Harry Kingston.

The astounding commercial success of STAR WARS caused every movie studio to attempt to re-package (in some cases re-re-re-package) its properties and BUCK ROGERS is no exception. The January 20, 1982 VARIETY had a two page add proclaiming BUCK ROGERS, HE'S BIGGER THAN EVER, in 2½ inch computer typeface. The ad copy reads:

> BUCK ROGERS.
> THE LEGEND.
> THE ORIGINAL SPACE HERO.
> THE FACE THAT LAUNCHED A THOUSAND ROCKET SHIPS.
> THE NAME PRE-SOLD TO MILLIONS OF FANS OF ALL AGES.
> THE SERIES THAT DELIVERED ONE OF THE HIGHEST CONCENTRATIONS OF THE UNDER-50 AUDIENCE IN ALL OF TELEVISION.
> BUCK ROGERS
> 37 HOURS OF FASCINATING INTERPLANETARY ADVENTURES, INCLUDING SIX TWO-HOUR FEATURES.
> AMONG THEM - THE BOX OFFICE HIT - BUCK ROGERS IN THE 25TH CENTURY.
> BUCK ROGERS. HE'LL PUT YOUR STATION'S RATINGS INTO ORBIT!
> AVAILABLE NOW FROM
> MCA TV

An earlier article in VARIETY explained that Universal was going to attempt to salvage the failure of their late 1970's Buck Rogers movie and TV series flops. The pitiful looking gun Buck is pointing at the readers of VARIETY (presumably the programmers for television stations) lacks either the hi-tech believable realism of STAR WARS' weapons or the naive charm of the weapons of the original strips. It's so simple and unimaginative our kids wouldn't play with it for 2 minutes. The new Buck himself in his tasteful, immaculate white suit looks like he just came from a Beverly Hills sportswear shop and is about to go jogging along the beach in Malibu - not to blast off for distant battles of nerve and wit with Killer Kane. Strader, taking a break between SESAME STREET and THE ELECTRIC COMPANY, just wandered in. He didn't like the gun either but he wanted the robot seen behind Buck. The robot was, in our opinion, almost as mediocre as the pistol. Never trust the esthetic judgment of space fanatics, especially one not quite 4 years old.

FLASH GORDON BILLFOLD. This vinyl wallet shows Buck's rival spaceman in a hurry. Flash did in fact do an admirable job of catching up with Buck in comic readership.

FLASH GORDON

Buck Rogers was born of Gernsback's pulp science fiction and the strip retained a lot of the futuristic, gadget-y flavor of that origin. FLASH GORDON published first in 1934 specifically came about ordered up by King Features to battle National Newspaper Services successful BUCK ROGERS strip but the character of the strip is entirely due to the unique talents of Alex Raymond, its creator. Somewhat inspired by the Philip Wylie/Edwin Balmer novel, WHEN WORLDS COLLIDE, the story concerns efforts by three people to prevent the destruction of the earth by the planet Mongo. The trio standing between us and destruction are Flash Gordon, Yale graduate and polo player, Dale Arden, his girl, and the useful scientist, Dr. Hans Zarkin, who has of course built a spaceship. Once on Mongo they find the not so nice "Ming the Merciless" running things. True-blue idealist that he is, Flash in no time begins to create discontent among the oppressed locals and by 1941 he has liberated Mongo. What gave the work success was Raymond's elegant brush work (rarely did he use a pen) and his complete grasp of the Burroughsian fantasy method.

Although a science fiction strip it proceeds with elementary relations - personal courage, love, loyalty, loathsome evil and cruelty. We see vast, awesome, mystic landscapes, bizarre, ruined cities and the exposed backs of bound beautiful women being mistreated. Flash may have used high technology to get to Mongo but once there he uses swords as much as ray guns to protect his female. He is a polished, worldly swashbuckler. Buck is the kid next door with a fixed up jalopy who becomes a fighter pilot. Flash is Yale: Buck is State College. Dale, Flash's woman is beautiful with long, flowing tresses. Wilma, Buck's girl, is cute originally with the short hairdo of a stenographer.

Alex Raymond died in 1956 in an automobile accident. He was at the height of his career which included the creation of 4 major strips - FLASH GORDON, JUNGLE JIM, SECRET AGENT X-9, and RIP KIRBY. The admirers of FLASH GORDON are many and diverse. Federico Fellini, the Italian film director, describes the effect of FLASH GORDON on him: "..a real hero even if his achievements were in a remote and fantastic world. I had a profound affection for FLASH GORDON and his creator. When I think of it, it seems as though he actually existed. At times in my films, I seek to find the color and verve of FLASH GORDON and his world." George Lucas tried unsuccessfully to obtain the rights to FLASH GORDON before he created STAR WARS. FLASH GORDON, as we said, made it to the screen before BUCK ROGERS. Flash is all action and glitter - perfect for the silver screen.

Stellar as his character is, collectors have relatively few Flash Gordon toys and memorabilia to acquire compared to Buck Rogers. Perhaps it was unaggressive merchandising, perhaps it lay in the less technological approach of the strip. Items have come out however through the various revivals of the character.

The concept was so well worked out that the strip enjoyed continuity even after its creator left it. When Alex Raymond joined the Marines in 1944 the strip was taken over by Austin Briggs, his assistant. Briggs left in 1948 and Mac Raboy drew it until 1967 when Dan Barry took over, assisted by his brother, Al Williamson, Ray Krenkal, Ric Estrada and Frank Frazetta (who also worked on BUCK ROGERS comic books at times).

As in the case with BUCK ROGERS a number of comic books published by King, Charlton and Harvey have been made through the years. Most are reprints from the newspaper strip but some original work has been done by Al Williamson, Reed Crandall, John Lehti and Paul Norris. To work in Raymond's shadow is, among cartoonists, an honor. Nostalgia Press has published two hard bound collections - FLASH GORDON, IN THE ICE KINGDOM OF MONGO (1967) and FLASH GORDON, INTO THE WATER WORLD OF MONGO (1971).

Flash had a 1930's and 40's radio program as well as Buck. The 1936 movie serial starring Buster Crabbe was mentioned. In it Charles Middleton was Ming the Merciless. Crabbe and Middleton fought it out in space in two more excellent serials FLASH GORDON'S TRIP TO MARS (1938) and FLASH GORDON CONQUERS THE UNIVERSE (1940).

A less than memorable TV kids' series was made of FLASH GORDON in 1953-54. Like BUCK ROGERS it could not scale itself down to the limitations of live television as well as CAPTAIN VIDEO, SPACE PATROL and other shows indigenous to the medium. In the 1950's when the recognition of FLASH GORDON was at ebb his name was cut from the old Buster Crabbe serials titles and such things as SPACE SOLDIERS TRIPS TO MARS and SPACE SOLDIERS CONQUER THE UNIVERSE appeared in syndication. Truly TV knows no gratitude. Even if you conquer the universe your name can get dropped when the ratings go and the public recognition factor isn't there any longer.

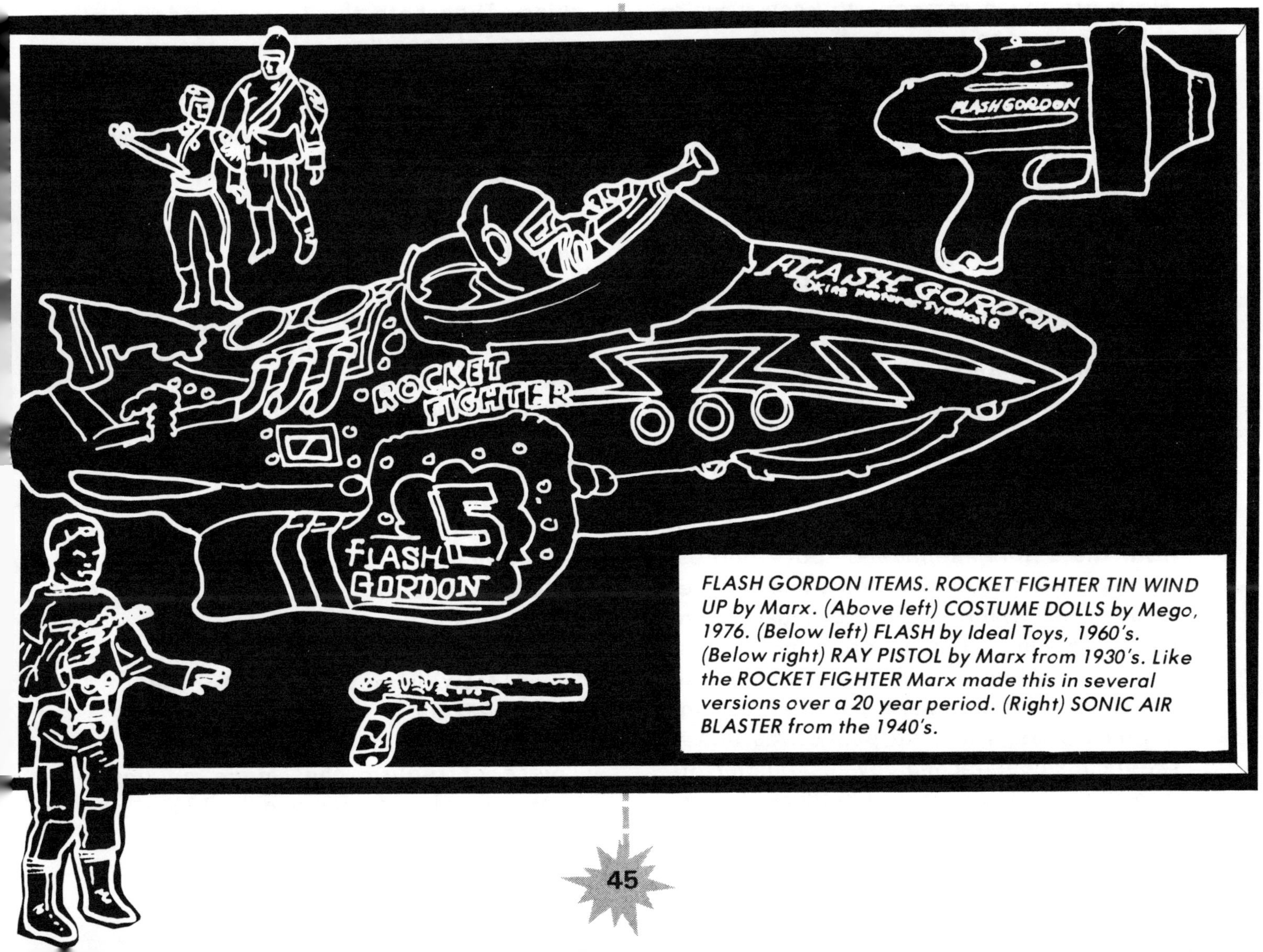

FLASH GORDON ITEMS. ROCKET FIGHTER TIN WIND UP by Marx. (Above left) COSTUME DOLLS by Mego, 1976. (Below left) FLASH by Ideal Toys, 1960's. (Below right) RAY PISTOL by Marx from 1930's. Like the ROCKET FIGHTER Marx made this in several versions over a 20 year period. (Right) SONIC AIR BLASTER from the 1940's.

BRICK BRADFORD

Another 1930's space comic strip was BRICK BRADFORD written by William Ritt and drawn by Clarence Grey. Brick first opened in 1933 as a daily and in 1934 as a Sunday page. Although it had a very long run (now being done by Paul Norris) and spawned comic books and a 1948 movie we haven't seen many Brick Bradford toys or memorabilia. Probably some more exist but they can't be very abundant. Brick lacked Flash Gordon's verve and Buck Rogers' technological genius so perhaps he did not lend himself well to spinoffs. It was action adventure with the hero skipping into the future and the past via time machine but was not particularly imaginative. Buck and Flash are tough company to keep if we're thinking "recognition factor" - which is how toy makers seeking licensing privileges look at things.

SUPERMAN

An alien who comes to the earth in a spaceship from a doomed planet would seem to be an appropriate figure to reckon with in a book on science fiction toys. Superman is from outer space and lots of toys have been made with his name or image. Yet the man of steel seems light years away from Buck Rogers and Flash Gordon when collectors think space or science fiction toys. Why? Superman though produced by science fiction writers and editors has a mythological dimension that separates him from the before mentioned space heroes. His meteoric rise to fame came about largely because of the public's disillusionment with real life role models (politicians, bankers, captains of industry and of course out of work fathers) during the Depression. It was a time when the public was either skeptical about technology or couldn't afford the benefits of it. Unlike Buck, Superman didn't need elaborate weapons or gadgets to combat evil. What Superman couldn't do with his physical strength wasn't worth doing. Superman's entire investment in hardware was a cape and a pair of tights. This fairly well eliminated him as an inspiration for great space toys which are in the final analysis little celebrations of technology. Superman is faith in the supernatural. Buck Rogers is confidence in science.

1950'S SPACE COMIC STRIPS

Public interest in rockets and space exploration was high in the post-war world. Yet fictional accounts of space were soon to encounter killing competition from the real thing. Then both the fantasy and realistic modes of popular science fiction were generally received with lukewarm commercial success. Horror did a little better job at capturing the imaginations of a public now alienated by atomic weapons and confused about their place in the world remade by modern technology - the space program included. The following 1950's newspaper strips didn't last long. Mostly they were realistic action adventures. If they resulted in toys or character collectibles we haven't seen any.

CHRIS WELKIN - PLANETEER

By Rus Winterbothan and Art Samson.

JET SCOTT

By Sheldon Stark and Jerry Robinson. Its hero was a scientist who worked for the Pentagon.

BEYOND MARS

Written and drawn by Lee Elias.

TWIN EARTHS

By Al McWilliams and Oscar Lebeck

DRIFT MARLO

By Tom Cooke and Phil Evans was a '50's piece of realism that made an unsuccessful try for life again in the 1960's.

SPACE COMIC BOOKS THAT NEVER GOT OFF THE LAUNCH PAD

Numerous comic books echo the dilemma of post World War II pop science fiction. Neither tradition - the Burroughs fantasy or the Gernsback technology clicked - as witness these short lived comic book efforts.

ROCKET KELLY, 1944-1945
ROCKY JONES, SPACE RANGER, 1955
FLYING SAUCER, 1950.
LARS OF MARS, 1951
MOON GIRL, 1947-1949
SPACE ACTION, 1952
SPACE BUSTERS, 1952
SPACE DETECTIVE, 1951-1952
SPACE SQUADRON, 1951-1952
SPACE WESTERN, 1952-1953

JAPANESE SCIENCE FICTION COMICS

The Japanese had a cartoon magazine by 1905. By the 1920's they were producing regularly scheduled comic books in color - 10 years before the Americans. So instantly successful was the Japanese comic book industry that lending libraries exclusively stocking comic books were founded. In order to supply the insatiable demand the lending libraries produced original material by 1930. It may seem strange to consider the popularity there of a form of communication we may have thought characteristically American but the Japanese have had a powerful tradition of drawing that lends itself to the form. Japanese woodcuts and brush drawing hundreds of years old have a simplicity, dynamic compositions, and expressiveness that many modern cartoonists would do well to achieve.

This is not to say that American comics were not important to the origins and development of Japanese comics. The 1905 Tozoko Puck was founded by Rakutan Kitazewa whose early work was heavily influenced by Outcaults and Opper.

The excellence of the Japanese space toys may have some relationship to the popularity of the science fiction comics there after World War II.

PUTCHA IN WONDER WORLD

In 1947 FUSHIGINA KUNI NO PUTCHA (Putcha in Wonder World) by Fukijiro Yokoi was first published. The hero of the strip, Putcha, was the son of a scientist. With his robot companion Perii, and Dr. Banbaran, the ever useful scientist/inventor Putcha travels through the universe of 2047. Among other things the trio discover the element X-nium which can neutralize atomic weapons. This would be, as one can imagine, a popular idea in post war Japan. It was a big hit in its time and began a wave of popular science fiction interest. The comic was continued by Tetsua Ogawa after the death of Fukijiro in 1948.

JAPANESE SHOGUN/ROBOT HEROES. The hybridization of myth and science fiction has produced a remarkable body of comics, television programs and toys in the Far East. Time and again in Japanese science fiction we see the blurred distinction between men, animals and machines due to their philosophical and religious heritage. Oriental religions seem not to place man apart from the rest of nature as do the Judeo-Christian traditions of the West. Even robots are frequently accepted in Japanese comics as full fledged sentient beings.

SPACE COMICS. A collection of comic books relating to science fiction provides the toy collector with examples of robot and spacecraft design that toy designers have often borrowed from. Familiarity with the history of science fiction movies and TV shows is helpful in understanding the social context in which certain kinds of toys were originally marketed. It's sometimes hard to see the old movies or TV programs so comics based on these can provide some idea of what the story was about. We have found comics relating to science fiction and space generally reasonably priced — excepting of course 1930's and 1940's BUCK ROGERS and FLASH GORDON's.

TETSUWAN-ATOM

Called "Astroboy" in U.S. TV animated cartoon releases, TETSUWAN-ATOM was created by Osamu Tezuka in strip form in 1951. Originally it was titled ATOM TAISHI. Osamu, who was quoted in the chapter FACE OF THE MACHINE, was born in 1926. He created his first science fiction feature in 1948. In 1961 he founded Mushi Productions which produced TETSUWAN-ATOM as the first series of animated cartoons for television in Japan.

It is a unique strip from a Western viewpoint. The hero, Atom, is a robot created by a scientist, Dr. Tenma to replace Tenma's own son, killed in a traffic accident. Atom does not grow as time goes by and Dr. Tenma comes to view him with antipathy, selling him to a circus. Finally the boy-like robot is acquired by kindly Dr. Ochanomuzu who inspires Atom to lead a life in the service of humanity. In the 1951 to 1968 run, the little robot boy battles Hot-dog soldiers, which are cyborgs with the brains of dogs, and several evil giant robots. TETSUWAN-ATOM seems a parable of the Japanese belief and confidence in (even friendship with) the machine.

TETSUJIN 28 GO

A comic strip competitor with TETSUWAN-ATOM was TETSUJIN 28 GO (Iron Man No. 28) created by Mitsuteru Yokoyama in 1958. Like TETSUWAN-ATOM it featured a robot. In this strip Dr. Kanada and Dr. Shikishima created a race of robots to use in World War II (called the Pacific War by the Japanese). The U.S. Air Force bombed the robots but Drs. Kanada and Shikishima kept the blueprints and in 1955 they produced TETSUJIN 28 GO for peaceful purposes. Controlled by the son of Dr. Kanada, Tetsujin was a metallic giant with helmeted head and a rocket mounted on his back. The popularity of this strip produced a radio program and a TV cartoon series (shown in the U.S. as GIGANTOR). Other Japanese strips that took the giant robot theme were THE DAI MACHINE, GIANT ROBOT and KING ROBOT. No wonder the Japanese population now welcomes robot- run factories. The present generation of workers grew up watching science fiction cartoons full of heroic robot stars.

CYBORG 009

This strip by Shotaro Ishimori began in 1964. It featured not a robot but a cybernetic organism ("cyborg" for short), a rebuilt human with some improved electronic or mechanical parts. Originally Joe Shinmamura, a juvenile delinquent, CYBORG 009 was the result of experiments by the evil Black Ghost attempting to create the perfect soldier. Fortunately CYBORG 009 (and his compatriots Cyborgs 001 through 008) did not turn out to share the leader of the Merchants of Death's rotten ideas and they in fact defeated him and ultimately many more monsters, evil robots and even the gods, until the strip finished in 1970. It was a TV cartoon as well.

AMERICAN POP CULTURE AS UNCHARTED TERRITORY. Looking back through 20 year old LIFE magazines and contemplating children's toys such as this BUCK ROGERS ATOMIC PISTOL by Daisy and Marx's BUCK ROGERS ROCKET POLICE PATROL SPACESHIP we may become aware that there are areas of our past that are not as we thought they were at the time. A surreal sensation comes over us. What was familiar is alien, what was reliable is foreign, what was sensible is mad. Reviewing yesterday's popular magazines we see how the public flies through uncharted, factless galaxies in mediacraft propelled by hype, only to land (and sink without a trace) on planets of fluff. The comics and toys by contrast may be equally naive but their sincerity redeems them.

RADIO SPACE OPERA

OW THE MARTIAN "INVASION" UNCOVERED
HE MASS HYSTERIA OF A GREAT NATION

WHEN on the evening of October 30, 1938, a young man named Orson Welles sent a wave of hysteria sweeping across the country by broadcasting a fictional news account of an invasion of U. S. by giant men from Mars, thousands of American citizens fled from their homes in absolute panic; police stations were swamped with 'phone calls; and patients in hospitals leaped from their beds seeking escape from approaching extermination. An imaginary interplanetary invasion had caused one of the greatest mass delusions in history. Eminent psychologists, in attempting to explain the phenomenon of how radio fans could listen to four announcements during the broadcast explaining that the fantasy was based on H. G. Wells's "The War of the Worlds" and still become panic-stricken, offer varied opinions as to the cause. Most of them agree that since radio entertainment is based on emotionalism, most listeners don't get beyond that stage—they don't stop to think. That is why the airwaves have become powerful instruments in the hands of dictators. The very effectiveness of radio has put democratic nations on the defensive

International

AT 23, Orson Welles (arms upraised), guiding genius of the "Mercury Theatre of the Air," has had more than a year of scaring children via the ether. His Martian thriller got out of hand when it terrorized a nation.

RADIO MAGIC. Juvenile science fiction was successful in the early days of radio. Special effects were cheaper in sound than on film and until TV came on the scene radio space operas derived from comic book characters attracted a loyal following.

ORSON WELLES' RADIO BROADCAST OF WAR OF THE WORLDS. The realism achieved in the golden era of radio was too intense for Depression era listeners. Orson Welles' radio drama of October 30, 1938, adapted from H.G. Wells science fiction classic about an alien invasion of Earth, caused pandemonium.

It is hard to imagine the magical effect of early radio on the public. Machines had talked before - but the cylinder and then the disc phonographs repeated what you already knew was there. The radio medium was mysterious and exciting. At first no one knew what to use public access wireless for except for announcements of public events or the news. But soon comedians, dramatists and musicians realized the power of voices and melodies issuing from a box. Programs with a fixed cost, presented on a regular basis and with characters the audience identified with were attractive early on to commercial sponsors whose products were targeted for that group. "Soap operas" flowered in the early days of radio, aimed at housewives and sponsored, as the nickname indicates, primarily by manufacturers of household cleaning products. Not long after serial radio programming intended for the younger set began. Cereal companies were the dominant advertisers.

With similar repetitive elements, main characters psychologically not unlike the audience, and strong fantasy overtones kiddie serials quickly seized and held loyal audiences. Children's radio shows created several habits that relate to space toys. "Space Operas," BUCK ROGERS and FLASH GORDON, were a successful radio genre as were "horse operas" like THE LONE RANGER and ROY ROGERS. They fulfilled the emotional needs of the audience without exception. All the radio shows generated a wealth of plastic, paper and pot metal trivia known as premiums. BUCK ROGERS premiums alone number more than 30. The patterns established by radio space operas carried over into television and to some extent into movies. STAR WARS is the perfected logical development of the space opera form established by comics and continued in radio and television. But kids today can't get the STAR WARS toys for a box top and a quarter. A complete set of Kenner's STAR WARS/EMPIRE STRIKES BACK figures, spaceships and play sets would have bought several Buicks back in the BUCK ROGERS generation. Inflation, like Lucas' titles, seems to stretch before us from infinity into eternity.

BUCK ROGERS IN THE 25TH CENTURY RADIO PREMIUMS. A galaxy of toys, pictures, badges and other paraphernalia was offered in the 1930's and 1940's to induce boys and girls to eat Cocomalt, Cream of Wheat and various Kellogg's products. Most were available for a combination of coins and box tops or coupons that came with the cereal.

TELEVISION SPACE OPERA

FLICKERING MEMORIES. What magic. Through the gray electronic snowstorm came the 1950's space heroes with yearly budgets that wouldn't buy a half a day's filming of STAR WARS. They were naive programs for a naive audience but we remember them with a smile.

IMPROVED RECEPTION, BETTER PROPS. The TV space operas of the 1960's were better made but with the exception of STAR TREK they have not faired well as pleasant memories. In some the robots were the most memorable characters. Mass audiences were becoming sophisticated due to media coverage of the real exploration of space. It would take George Lucas' STAR WARS of the late 1970's to credibly revive the space opera.

Mass audience science fiction T.V. series began with CAPTAIN VIDEO in 1949 and have remained only juvenile interest or special interest programming ever since. Head on against situation comedies or cowboy or cop shows, space or science fiction programs can't compete in prime time. Even STAR TREK, considered to be among the most popular TV space programs, was a commercial failure in its 3 years of prime time exposure. Only in syndication did it achieve an appropriate balance between its air costs and its ratings. Rod Serling's TWILIGHT ZONE, among the finest of science fiction TV shows, was to be dropped after its first season. Only the 2,500 letters from fans in support of it saved it. TWILIGHT ZONE went on another five seasons but never made it into the top 20 shows then. Today it and STAR TREK are playing somewhere every day in syndication. We can conclude that space and science fiction fare are inherently specialized programming and with the advent of cable TV allowing dozens of channels it may prosper in reruns. New television programs with a space theme may be handicapped in the fulfillment of this specialized audience's needs by reruns, the present day inflated costs of new productions and the high technical standards set by the movie STAR WARS and its sequels.

The first space TV program was CAPTAIN VIDEO. Its popularity begat TOM CORBETT SPACE CADET, SPACE PATROL and others now forgotten like ROCKY JONES SPACE RANGER and ROD BROWN OF THE ROCKET RANGERS. They basically took the concept of the BUCK ROGERS radio shows with unmarried red blooded American heros, likable side kicks, plucky girl companions and villians of the boo-hiss category. In terms of content they were not dissimilar to the numerous cowboy kiddie shows. TV didn't generate as many premiums as the radio shows but the 1950's space shows were an important bridge in the popular perception of space between the 1930's fantasies of BUCK ROGERS and FLASH GORDON and the 1960's realities of Mercury and Apollo.

CAPTAIN VIDEO

We hear Richard Wagner's "Flying Dutchman" overture and the stern baritone announces: "Captain Video! Master of Space! Hero of Science! Captain of the Video Rangers! Operating from his secret mountain headquarters on the planet Earth, Captain Video rallies men of good will and leads them against the forces of evil everywhere! As he rockets from planet to planet, let us follow the champion of justice, truth and freedom through the universe!" All this the able space commander did on a weekly prop budget of $25. (If only Captain Video would work on the Federal budget today.) Though he talked of the wonders of space flight faster than the speed of light the Rangers tooled around in a cardboard space ship with the dials and instruments clearly painted on. His "Opticon Scillometer," which could see through walls, was a piece of pipe. In the early programs his Rangers were often represented by clips from ancient cowboy movies. In spite of the pitiful props and skimpy budget CAPTAIN VIDEO was fresh and like so many of the pioneer television programs it is warmly remembered. Not all the reasons to recall this first of all TV space operas are nostalgic or sentimental. The series had charm and originality. Until the DuMont network went under CAPTAIN VIDEO was a bright spot in early children's television programming.

For fans whose memory may be fuzzy CAPTAIN VIDEO was first played by Richard Coogan and in 1951 Al Hodge (once the Green Hornet on radio) took over at the cardboard controls of the spaceship Galaxy. The Ranger, his young sidekick, was Don Hastings and the nefarious scientific genius, Dr. Pauli, was played by Hal Conklin wearing a double breasted suit and dangling a cigarette in a holder - was a sort of cross between a ganster and a Nazi, as Donald F. Glut has described him in his and Jim Harmon's authoritative and delightful book, THE GREAT TELEVISION HEROES. The much used and re-used special effects were by Russell and Haberstroh and were various and imaginative considering the paltry budget and the limitations of live TV (although the special effects were filmed). The show was created and produced by James Caddigan and written by Maurice C. Brock. Like Buck Rogers, the Captain had a technocrat's trust in the latest invention. Besides the Opticon Scillometer already mentioned he utilized a Remote Tele- Carrier or the Mango Radar Screen to spy on enemies of peace, and a Cosmic Ray Vibrator to shake sense into evil doers and in the cases of last resort an Atomic Rifle, although the Captain was always reluctant to take a life. One of the reasons Captain Video didn't often kill the bad guys was that fans often developed an affection for them and demanded their return after being eliminated. Dr. Pauli made a miraculous reappearance after being killed four times. Even when blown to bits the Captain was forced by fan mail to reassemble him (with the help of a fabulous invention) atom by atom.

Dr. Pauli by the way was no slouch in the technology race. He invented a cloak of invisibility, a barrier of silence and the remarkable Trisonic compensator which could curve bullets around corners.

Among the 300 other fiends the Captain and his Rangers had to cope with in the 6 year run were Kul of Eos, Dr. Clysmok,

Dahownie, Nargola, Heng Foo Seeng and Mook the Moon Man.

Most memorable of the sequences concerned the giant robot Tobor ("robot" spelled backwards) who, though designed to be a useful worker, was reprogrammed by Atar, a very, very bad girl-type. Tobor was nothing to be taken lightly. From the chest of his cylindrical body of indestructible metal was a death ray gun. He sported rockets on his back which enabled him to pursue the Galaxy through space. At the end of his arms were great hooks. After numerous encounters in which Tobor seemed destined to extinguish good from the universe the technical genius of the Captain finally triumphed. Making a transmitter that gave orders on the same wave length as Atar, his evil mistress, Tobor was torn apart when the contradictory orders began coming in. He self-destructed in a systems overload. Again, as in the case of Dr. Pauli, fans wanted him back. Retrieved from an astroid the Captain remade Tobor's controls to respond only to him. Then Tobor battled a duplicate made from his own plans and continued working for good in the universe - occasionally doing a guest spot on CAPTAIN VIDEO.

A particularly significant appearance of Tobor and one that illustrated how good the show could be concerned Tobor being sent to a planet with a poisonous atmosphere. The Captain had devised a method of projecting his own mind into Tobor's head allowing him to experience an environment so hostile that even a space suit could not protect him against it. In a profound metaphor of the interdependence of technological man and his machines Captain Video, sitting in a chair on Earth in his secret headquarters, felt what Tobor experienced.

Columbia Pictures made a 15 part movie serial called CAPTAIN VIDEO produced by Sam Katzmann and directed by Spencer G. Bennet. The television cast was not in it and although the film was better funded it did not cut it with the kids. Judd Holdren played Captain Video and Larry Stewart was the Ranger in the movie serial. Hollywood never seems to learn that props and special effects alone do not a popular science fiction work make.

TOM CORBETT, SPACE CADET

Adapted from a juvenile novel by the respected science fiction writer, Robert A. Heinlein, and with German rocket expert Willy Ley as technical advisor, this early 50's TV series was more authentic in its details than CAPTAIN VIDEO but no less delightful. Young Tom (played by 29 year old movie actor Frankie Thomas) was enrolled at the Space Academy. He and his wise cracking buddy, Roger Manning (played by Jan Merlin), and their fellow cadets would periodically leave the Space Academy and take the spaceship Polaris out for a spin into outer space. Skipper of the Polaris was Captain Strong (first played by Michael Harvey, later by Edward Bryce) and the ship's Venusian navigator was Al Merlin. Captain Strong's girl friend was Dr. Joan Dale (Margaret Garland). This jolly crew encountered its share of space pirates, dinosaurs and weird aliens but they also had to dodge anti-matter, meteors and steer around asteroid belts between Mars and Jupiter. Weightlessness in space was authentically depicted. Thanks to Mr. Ley the show had the ring of the real space age to come from the very beginning of the program when over a montage of shots of the academy and a real V-2 rocket blasting off the announcer would say: "Blast off minus five...four...three...two...one...ZERO! As roaring rockets blast off to distant planets and far-flung stars, we take you to the age of the conquest of space...with TOM CORBETT, SPACE CADET!!"

The Polaris itself was a V-2 sort of craft with round portholes and side fins. What else would one expect from one of the pioneers of German rocket research.

TOM CORBETT was first telecast on October 2, 1950. It was a prime time show during 1952 and then moved to Saturday daytime where it continued off and on until 1955. A radio version featuring the TV cast was also heard on ABC.

SPACE PATROL

Beginning as a local afternoon serial on California station KECA, SPACE PATROL was picked up by ABC for a Saturday morning spot. It was like CAPTAIN VIDEO and TOM CORBETT, a live juvenile space opera that generated considerable premiums and character toys. In terms of scientific realism it was in between the technically correct TOM CORBETT series and the fanciful CAPTAIN VIDEO. In terms of stalwart defense of interplanetary justice, SPACE PATROL's Commander-in-Chief, Buzzy Corry, was second to none. Buzzy came from a space oriented family - his father was a rocket designer. The man who played Buzzy Corry was Ed Kemmer, a World War II pilot who learned acting when he was captured and imprisoned by the Nazis after being downed in Germany.

The creator of the series, Mike Moser, said he dreamed up SPACE PATROL in a Navy plane flying over the Pacific. He was an unabashed believer in technology and his concept for the series was to promote the public's interest in and acceptance of the wonders of science. It was a sincere and effective effort. For a budget of $2,500 per episode 50 people gave their best to make the show popular with the young viewers.

The special effects were done by Paul, Oscar and Franz Dallons, former medical research designers who also did the special effects for movies like DONOVAN'S BRAIN. They frequently redesigned and improved the Terra, the Patrol's space craft. Terra I was an old fashioned Buck Rogers vintage horizontally landing rocketship and upon objections from the fans they began vertical landings and take offs to conform to the V-2 mode the kids would see on the news.

By 1955 when the show ended, Commander Corry was blasting about in the improved Terra V. Possibly that implanted the need to trade for Detroit's latest each year in the impressionable viewers. At any rate the not so subtle underlying moral of all the 1950's space operas was an endorsement of the materialistic, technologically oriented mainstream American commercial culture of the times. They reached an audience that had never heard of drugs, women's liberation, or pollution. Consider the ethics of the Junior Rocket Rangers - a fan club of ROD BROWN OF THE ROCKET RANGERS, a short lived 1950's space television opera.

On my Honor as a Rocket Ranger, I pledge that:

1. I shall always chart my course according to the Constitution of the United States of America.
2. I shall never cross orbits with the Rights and Beliefs of others.
3. I shall blast at full space-speed to protect the Weak and Innocent.
4. I shall stay out of collision orbit with the laws of my State and Community.
5. I shall cruise in parallel orbit with my Parents and Teachers.
6. I shall not roar my rockets unwisely, and shall be Courteous at all times.
7. I shall keep my gyros steady and reactors burning by being Industrious and Thrifty.
8. I shall keep my scanner tuned to Learning and remain coupled to my Studies.
9. I shall keep my mind out of free-fall by being mentally alert.
10. I shall blast the meteors from the paths of other people by being Kind and Considerate.

Or the oath of the Video Rangers Fan Club:

> We, as Official Video Rangers, hereby promise to abide by the Ranger Code and to support forever the cause of Freedom, Truth and Justice throughout the universe.

A lot of Rocket Rangers and Space Cadets and Junior Video Rangers were to go to Vietnam to find an alien people enslaved and yet indifferent to offers of freedom and justice. And what of space operas' credibility when Neil Armstrong hopped out of Apollo 11 and Mook the Moon Man was nowhere to be seen on that desolate moonscape? The real spacecraft itself had been transformed into a tiny little bucket of bolts compared to the majestic, sleek, needle nosed crafts called Galaxy, Terra and Polaris. Concern about pollution, the anti- nuclear movement and growing mysticism and cynicism were to blacken the prospects for this kind of naive, optimistic science fiction from the mid-50's on. Until the brilliant synthesis of space opera and high technology called STAR WARS relit the rockets of popular imagination the genre was adrift. To be sure there were TV science fiction series but only STAR TREK had anything of the popular impact of these 1950's naive but charming space operas. Most of the rest are as empty and impersonal as outer space itself. They reached mass audiences and some generated space toys and collectibles so they deserve mention.

ROD BROWN OF THE ROCKET RANGERS

A 1953 space opera with Cliff Robertson as Rod Brown. Hokey and short lived it lasted only one season.

COMMANDO CODY

The Commando was Judd Holdren. He hosted a cheapie kiddie show that consisted largely of re-runs of old Republic serials.

ROCKY JONES, SPACE RANGER

Richard Crane starred. It was the first TV space show made completely on film.

CAPTAIN 2-RO

The gimmick of this one was a time machine which sent Roy Steffens as the Captain into the past and future. He wrote, produced and starred in it.

PHANTOM EMPIRE

This science fiction juvenile starred the singing cowboy, Gene Autry, who lived in a future city under the Earth.

LOST IN SPACE

Irwin Allen, the producer of the movies TOWERING INFERNO and POSEIDON ADVENTURE, produced a number of prime time TV series in the 1960's. These were acceptable as family entertainment but lacked both the magic of fantasy or the realism of true science fiction. Following the futuristic adventures of the crew of the glass nosed atomic submarine, Seaview, in VOYAGE TO THE BOTTOM OF THE SEA (1961), Allen took his Hollywood cliches into space.

A CBS Wednesday night family program, LOST IN SPACE was about Professor John Robinson, his wife and three children who embark upon a 5 year voyage of exploration of a planet in the Alpha Centauri star system and become lost in space due to Dr. Zachary Smith's sabotage of the control system. Smith, who works for a foreign government, is trapped in the space ship so a truce has to be worked out between him and the pilot, Major West, and the Robinsons. For 3 seasons they wandered from strange planet to strange planet encountering weird monsters and bizarre forms of alien life. Dr. Smith was true to his character always working at cross purposes with the rest of the group attempting various deals with aliens. Like the old serials the program often ended in a cliff hanger.

The helpful robot also doubled as part of the computer that controlled the ship. Like Robby of the movie FORBIDDEN PLANET and R2-D2 of STAR WARS, the unnamed LOST IN SPACE robot is endearing, obedient, kind, cheerful and works for free, never talks back or goes on strike. It is in the tradition of the perfect help, the kind you can't get anymore as they say.

LOST IN SPACE was first telecast on September 15, 1965 and aired its last episode on September 11, 1968. The principal characters were Guy Williams as Professor John Robinson, June Lockhart as his wife, Maureen, and Mark Goddard as Major Don West. The three children were: Judy Robinson played by Marta Kristen, Will Robinson by Billy Mumy and Penny Robinson by Angela Cartwright. The evil Dr. Smith was portrayed by Jonathan Harris.

Several space toys exist of the LOST IN SPACE family, the unnamed robot being the most abundant. It was more memorable than the cardboard space family.

Allen's next TV series was TIME TUNNEL a 1966-1967 effort which consisted of two handsome young scientists who get "lost in history" instead of LOST IN SPACE. Nothing Allen does is original with him and moreover he uses the same gimmicks time after time. In fact he loves cutting in footage from earlier movies. The TIME TUNNEL quickly became lost in the annals of TV GUIDE.

STAR TREK

The most influential space program on television during the years of actual space exploration was STAR TREK. It was the creation of Gene Roddenberry, a television writer knowledgeable about the history of science fiction. It achieved credibility because the focus of the programs was sociological science fiction - i.e. the implications of human (and sometimes alien) relationships in futuristic settings - instead of technological gimmickry which was hard to do credibly in a Southern California sound stage when real rockets were blasting off every few months from Cape Canaveral. Many of the episodes were in fact thinly veiled metaphors for the social problems of the 1960's - discrimination, the Cold War, even the tensions of corporate life seem reflected in various episodes as the starship, U.S.S. Enterprise, explores space and delivers supplies to distant Earth colonies. It was cancelled after its first year (1966) but unexpected audience support enabled it to cruise on two more seasons. Finally falling victim to poor ratings and a high percentage of young viewers who were unattractive to prime time advertisers, NBC axed it amid the outraged protests of the growing cult of fans called "Trekkies." Since then STAR TREK in syndication has made history. The networks put out an animated cartoon series (1973-1975) on Saturday mornings based on the series. A movie version was released in 1979. These periods of revival have created three groups of STAR TREK collectibles - those of the original TV show, the ones that accompanied the syndicated reruns and the items produced after the 1979 movie. Few are space toys in the strict

sense but because there is strong (but specialized) collector interest we are including a partial list of STAR TREK related licensed merchandise.

The cast of characters of the original TV show was: Capt. James T. Kirk (William Shatner), Mr. Spock (Leonard Nimoy), Dr. Leonard McCoy (DeForest Kelly), Yeoman Janice Rand (Grace Lee Whitney), Sulu (George Takei), Uhura (Nichelle Nichols), Engineer Montgomery Scott (James Doohan), Nurse Christine Chapel (Majel Barrett), Ensign Chekov (Walter Koenig).

BATTLESTAR GALACTICA

This was another TV space opera with some associated collectibles but one without much originality. It was first telecast on ABC Sunday night, September 17, 1978, and finished in 1979.

The special effects by John Dykstra (the technician on STAR WARS) were excellent

The frustrations of bad reception did not retard the growth of juvenile science fiction programming on television. As in radio space opera the continuity of characters (which were not difficult to understand) offered a certain reassurance to the audience analogous to the security which soap operas promised to housewives. The charm of TV space opera came from the mixture of the often pathetic sets and special effects sharply contrasting with the ambitious save-the-universe-from-absolute-evil goal of the heros. The poor sets and special effects came however to clash with the burgeoning space programs media releases and the naive idealism of the TV space heros ran against the tide of growing adolescent cynicism and worldliness as the '50's pushed into the '60's with drugs, riots and Vietnam. Captain Video, Tom Corbett and the others ran up against a force no space hero can conquer — the fickle, fleeting taste of the mass audience.

but critics dive bombed BATTLESTAR GALACTICA for its rip-off of STAR WARS.

After several pages noting the similarities between STAR WARS and BATTLESTAR GALACTICA Isaac Asimov sarcastically comments: "Get the subtle difference?" In the 7th millenium A.D. Galactica was the last battlestar after a treacherous attack by the evil cylons assisted by Count Baltar. Lasers flashed and smaller spaceships blew into atoms by the dozens. Both the TV series (which is said to have had a budget of one million dollars an hour) and the film version had more action than life. STAR WARS had lively characters in thrilling situations. BATTLESTAR GALACTICA had only situations. The STAR WARS producers sued ABC for "stealing" their ideas and audiences were indifferent.

The cast was: Commander Adama (Lorne Greene), Captain Apollo (Richard Hatch), Lt. Starbuck (Dick Benedict), Lt. Boomer (Herb Jefferson, Jr.), Athena (Maren Jensen), Flight Sgt. Jody (Tony Swartz), Boxey (Noah Hathaway), Col. Tigh (Terry Carter), Cassiopea (Laurette Spang), Count Baltar (John Colicos).

SPACE MOVIES

VOYAGE TO THE MOON. George Melies' 1902 film about space travel must have seemed implausible even at that early date. The audacity of the idea along with Melies' playful style made up for the absurdities and then as now boldness and novelty in the cinema produced commercial success. The bullet-like spaceship shown landing in the eye of the Man in the Moon is a concept widely adapted from this important early space movie.

MECHANIZED ROBOT. Several versions of Robby the Robot from the movie FORBIDDEN PLANET have been made. This 13½" tall, TN, Japan adaptation is like most of these — unlicensed and hence not named Robby.

Many of the earliest movies were science fiction inspired. The Frenchman George Melies, by profession a magician, exploited the deception and illusions the film medium permits in the early years of this century. Beautiful girls become demons before one's eyes, arms and legs are switched, nothing is secure or safe from change or metamorphosis. Like the subject matter of the magic lantern from which the concept of the cinema evolved in part, the movies began with presenting wondrous events, surrealist surprises, a vision of a completely modernist world. Jarring and shocking effects were easier to achieve technically in movies than poetic or even realistic effects - and vastly more rewarding. What could be more thrilling than a rocket trip beyond the earth? What setting is as intriguing and at the same time hostile and frightening as space? Melies' A TRIP TO THE MOON (1902) was but the first of a long line of space movies.

Problems of changing standards of credibility have plagued science fiction movies. While the camera can achieve wondrous deceptions, special effects acceptable to today's visually sophisticated film goers do not come easy or cheap. With a few notable exceptions the vast body of movies dealing with space cannot be watched with suspended disbelief. We see the wires; we find distracting space headgear that is too easily identified as football helmets with goggles. Even without the withering blasts of the real space program chilling the back lot fantasies of Hollywood, we encounter problems in taking seriously much of science fiction movie history. Along with the lack of adequate budgets for props and effects these low budget movies show deficiencies in script, direction and acting. Promotion and distribution were frequently less than stellar as well. Most science fiction films are B minus in the last analysis. At best they have a charm analogous to that of the early pulps. Except as 'camp' or 'kitsch' it's hard to enjoy them today. They tend to be gloomy, pessimistic, monster-filled allegories that teach an anti-scientific lesson. Even many of the better produced efforts warn of the pernicious effects of technology. Paul Wegener's 1914 GOLEM is a warning about artificial life as are of course all the Frankenstein films beginning with the first by J. Searle Dawley for Edison in 1910. Who has not seen a hundred times movies with mad scientists, murderous robots and mutated monsters? These cliches could once be thrown together in inexhaustible variety to allow cheap production costs for film makers and cheap thrills for audiences.

As space toys are by their nature a kind of advocacy of science and technology we must go to the science fiction films that are more positive or at least created public interest in the subject of space. We have already discussed the success of the FLASH GORDON movie serial of 1936 and the BUCK ROGERS movie serial of 1939. Realism (at least for that time period) was brought to the screen in 1950 in Universal- International's DESTINATION MOON, directed by Irving Pichel. Aliens in flying saucers came to the silver screen in 20th Century Fox's THE DAY THE EARTH STOOD STILL (1951), directed by Robert Wise, and THIS ISLAND EARTH (1954) a Universal- International film. The 50's were a time of suspicion, McCarthyism, and the Cold War. Since then we have been liberalized and psychoanalyzed as a culture and politically we repudiated McCarthyism and embraced detente. Is it any wonder that the aliens in Steven Spielberg's late 1970's film, CLOSE ENCOUNTERS OF THE THIRD KIND, are a lot nicer fellows than the 50's little green men? It remains to be seen which version will be closer to the truth if in fact we are not alone.

The actual exploration of space has produced no sign of even microscopic life in the solar system. Furthermore it has confirmed what scientists thought all along. Travel beyond the solar system involves almost insurmountable difficulties. Even if barely technically possible, the social, economic and biological costs may be judged too great for humanity to pay. With one exception - 2001, A SPACE ODYSSEY, by Stanley Kubrick - no film has tried to deal realistically with either the technical or psychological dimensions of the real space age. Realized in 1968 this movie, co-scripted with Arthur C. Clarke, is a milestone in space movies. The brilliance of this wide screen vision of space through which machines of great authenticity floated contrasted sharply with the enigmatic plot. Yet that contradiction created a feeling that many of us shared - we knew a great deal about space - the looks of rockets, the effects of weightlessness, the absolute blackness of the universe and yet we understood nothing. 2001 was a hugely successful movie at the box office and it was respected by the critics. It was the biggest budget science fiction film to that date. With its success Hollywood was willing to bet more on space movies. THE BLACK HOLE, BATTLESTAR GALACTICA and SILENT RUNNING had the high ticket special effects of 2001 but lacked that mixture of artistic originality and timeliness that characterizes a hit. Curiously all three were full of the dour moralizing tone of the old B-minus science fiction flicks. Only STAR WARS and its sequel, THE EMPIRE STRIKES BACK, proved to be winners in the post-2001 space movie race.

In an analysis of the frequency of appearance of key words in over 1,000 science fiction titles, one science fiction author found 'star' was the most used word by far. It was used 57 times. 'War' appeared as the 20th most used word, 5 times. In even title selection STAR WARS conjures up a lot of past science fiction. It breaks ground not so much in concept or content but in style. George Lucas' cinematic power has taken belabored stock characters derived from popular comics, serials, and movies and speeding through the space of his imagination they are polished into glossy objects that capture our eyes and hearts if not our intellect. STAR WARS is light years past all competition as the biggest space opera of all times. STAR WARS and its sequel, THE EMPIRE STRIKES BACK, are in fact the #1 and #2 biggest grossing movies ever. Only BUCK ROGERS of the 1930's has had comparable popular stature in its own period.

Heavy duty science fiction writers and intellectuals like Isaac Asimov liked STAR WARS. Why my 2 and 4 year old boys insisted on an all STAR WARS Christmas is beyond me when they have never seen either of the movies. Doubtlessly the TV commercials have something to do with it but toy motorcycles and plastic trucks are hyped on the tube just as much. The die cast and plastic replicas of the high tech space ships and various other fantastic methods of transportation have limited action, are a uniform dull grey and yet Strader and Ross play with them by the hour. Lucas' genius in capturing a deep collective subconscious of the times cannot be doubted. It permeates everything he and his organization touch.

Full to be sure of the sociological character of our times and even containing a bit of moralizing (Lucas' first film, THX 1138 - made in 1970 - was a low budget anti-utopian story set in the future) STAR WARS is a phenomenon to be reckoned with especially in a book on space toys. It was an unexpected hit as a movie and the response to the toys was even more startling. Kenner Products, a part of General Mills, won the bidding rights after the film had been released. Almost no toys had been manufactured when the Christmas of 1977 rolled around. Sears offered a Kenner-made cardboard Creature Cantina Action Playset with miniature action figures. By 1979 the sale of STAR WARS toys topped $200 million.

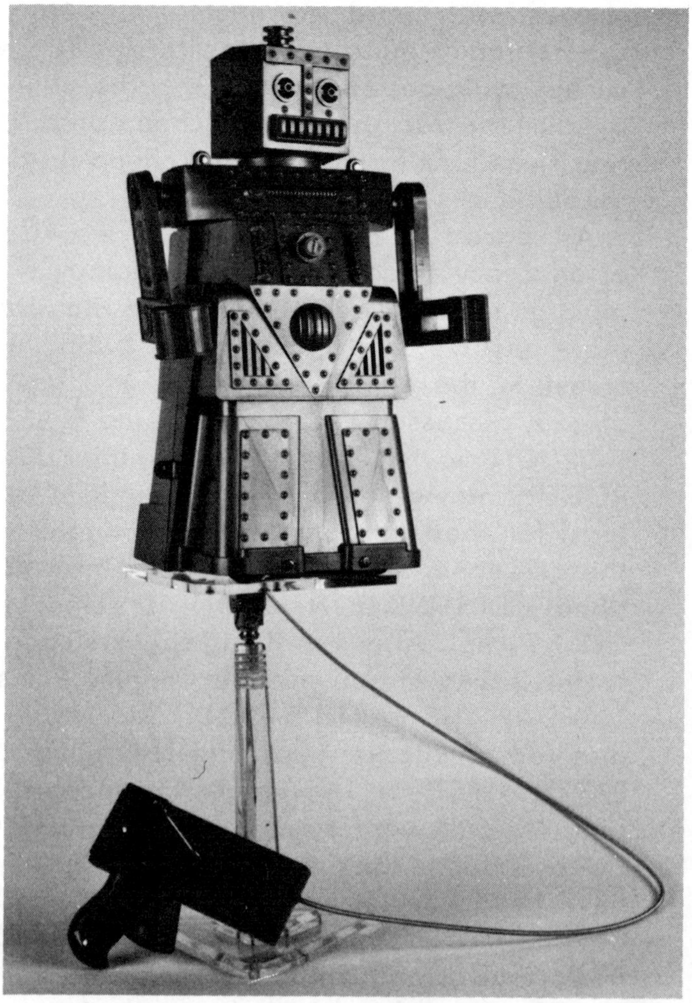

ROBERT THE WONDER TOY. The giant robot Tobor, star of the 1954 film TOBOR THE GREAT, had a pet robot, Robert. Ideal made a toy based on Robert that talks and can be remote controlled via the cable and pistol-like devise below.

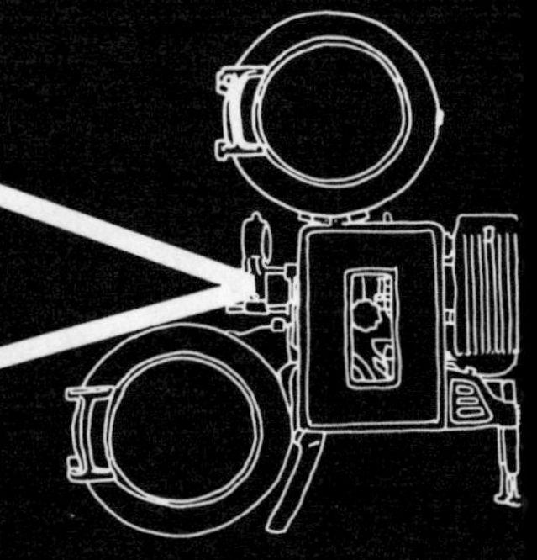

MYTH, MAGIC AND MOVIES

Apocalyptic visions are not new. The Book of Revelations in the Bible details a script that Hollywood's special effects wizards would go over budget on. Why humanity is attracted to stories of cosmic destruction with a cast of characters half of whom are good and the other half evil who never endingly battle each other requires pondering. Perhaps the early good box office of Revelations and similar epics in other cultures has created a domino effect. Each new generation is impacted by the rave reviews of the last. Perhaps there is a deep psychological reason. From the ancient myths to STAR WARS the saga of larger than life beings battling one another through the heavens has played uninterruptedly. Many of the world's cultures have similar visions of the end (or the beginning) of time.

The presentation of this story got a boost with the perfection of the magic lantern in the 1650's by Athanasius Kircher. The magic lantern used a lens that projected an image from a glass slide onto a wall. The earliest ones used a candle as a light source. Significantly it was first employed by magicians and entertainers who created a "phantasmagoria," a kind of shadowy ghost theatre. Gnomes, sorcerers and other mythic remnants of European pre-Christian times were the main actors. Frightening cosmic visions seemed perfect for the immateriality of an image made of light and shadow itself. It took several hundred years for ordinary people to let their portraits be projected by magic lantern, or slide projectors as we know them today.

First and foremost the magic lantern was an entertainment that dealt with the weird, the grotesque, the cosmic or an occasional historic figure who transcended mortality or landscapes of the most distant and awesome nature. When the magic lantern was combined with photography and the principles of motion illusion the cinema was born. As in the case of the magic lantern the immateriality of the projected picture made even more mysterious by movement found perfect subject matter in the dark mythic realm. Science fiction has important differences from myth but mass audiences do not distinguish structured supposition from fantasy. Consider the early science fiction movie successes: 1902 - A VOYAGE TO THE MOON, 1910 - FRANKENSTEIN, 1914 - THE GOLEM, 1922 - NOSFERATU, 1926 - METROPOLIS.

As long as magicians and story tellers, directors and special effects artists can keep updating Revelations and the cosmic battle for good and evil the lines at the box office will remain long.

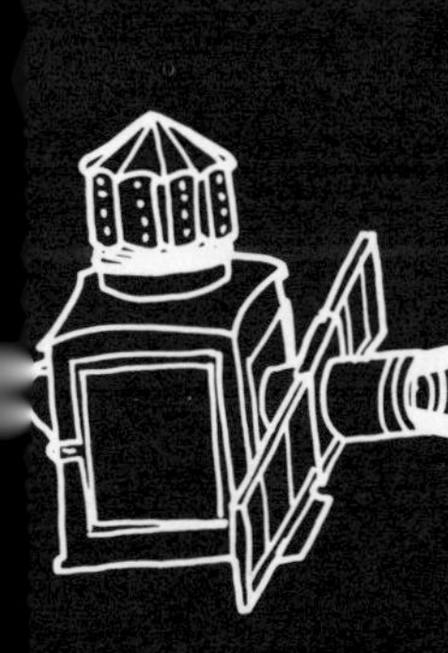

Francis Ford Coppola recently said "George Lucas owns a bank called STAR WARS." The saga of Luke Skywalker, Princess Leia, Darth Vader and Artoo-Detoo is planned to unfold in 12 separate films. Two have been released, STAR WARS in 1977 and THE EMPIRE STRIKES BACK in 1979. REVENGE OF THE JEDI is to be premiered in 1983. The amount of licensed material relating to STAR WARS is staggering at this point. If the remaining 10 parts of the STAR WARS epic are released at 3 year intervals it will be 2009 before peace comes to the universe.

We have heard that there are collectors attempting to obtain one of every STAR WARS licensed items. Such dedication requires several thousand dollars a year and ultimately a large warehouse. The outcome of such acquisitions as investment is problematic. All of us wish we could order some Mr. Atomics from the Sears 1962 catalogue. The conditions of manufacture are vastly different for the STAR WARS toys however.

Kenner is owned by General Mills, for example, and if cereal companies do anything well they know how to market a single product over a long time and wide area. In 1980 the General Mills Toy Group which includes Kenner, Parker Brothers, and Fundimensions, sold $674 million in toys. That's a few Monopoly sets and a lot of STAR WARS stuff. Bernard Loomis, president of the marketing and design service of General Mills says, "New merchandise is key, the way to keep properties alive is to bring out new themes and sub-themes each year." STAR WARS has half a hundred different figures "and there are kids whose mission in life is to own them all," comments Mr. Loomis. Because the next STAR WARS sequel is a year away Kenner will introduce a "dramatic new line" this year.

By contrast, the classic Japanese tin battery operated robots were produced by small companies and hand assembled in tiny workshops by subcontractors. Some models were made for only one season and small variations in color and paint are common. Before the early 1970's there were few collectors and no speculators saving them new in the box. Many of the 1950's and 1960's space toys are generally scarce if not rare.

While Kenner intends to introduce "new themes and sub-themes" they also seem intent on keeping alive most of the earlier versions. A review of their wholesale catalogues from 1978 to 1982 reveals that indeed most STAR WARS items are offered from year to year. Yet the popularity of the films and the fact that few will survive in pristine condition (most of the spacecraft have numerous small parts easily broken or lost) might make them some day important collectible space toys. For us there is no question of "to buy or not to buy" STAR WARS toys. Strader and Ross want one of everything and since they generously give us photo rights to their STAR WARS collection we may ultimately need to arrange credit with Kenner. We live with the "kids whose mission in life is to own them all" that Mr. Loomis referred to.

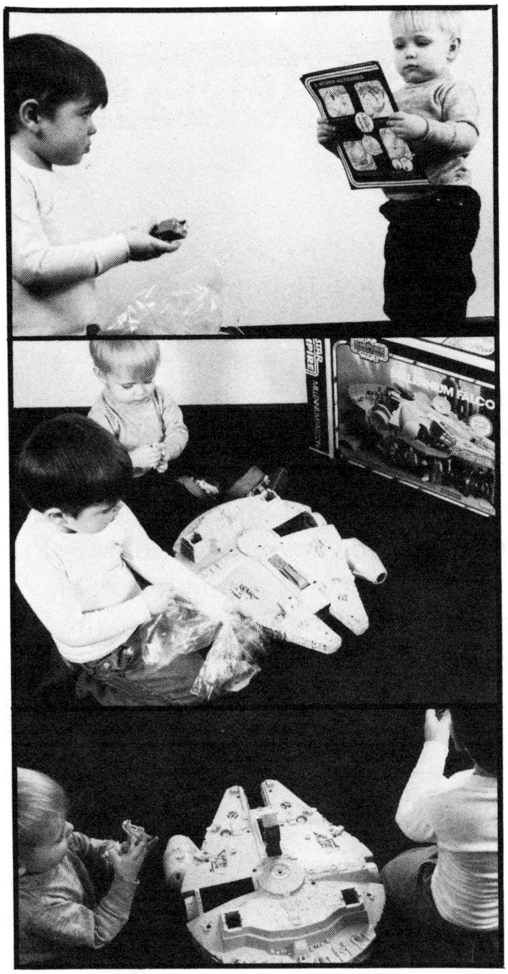

CHIEFS

SPACE TOYS FOR LOVE OR MONEY

To this point the book has related space toys to pop science fiction and to the space program. Perhaps now there has been an interest created or amplified or refined, defined or re-defined. Whatever - you want to buy some space toys or perhaps sell some. The subject of money is unavoidable. We have not printed prices in the book however. We do have available a Price Guide to the space toys in this book and some that are not. The last page in the book contains an order form if the book dealer you got this from was unable to supply you. It costs an extra $3.50 postpaid and the profit motive was not behind the decision to handle prices this way. Every year we will be revising and adding to the separate price guide as new information comes to us. Reprinting the whole book wasn't the answer. The need for current price information will be apparent when you consider the state of the market for space toys.

Eric Greenleaf wrote a brilliant article in the March, 1982, MAINE ANTIQUE DIGEST called "Antiques and the Product Life Cycle." In this perceptive piece he borrows from commerce the concept of product life cycle which he defines as "consumer acceptance a product goes through, from the time it is introduced to the time it is no longer saleable" and applies that concept to antiques and collectibles. Briefly the five phases of an antique or collectible are:

1. Junk Phase: When most people discard an object. The few weirdoes (or perhaps geniuses) who covet items of such low esteem must root around in yard sales and household auctions for junk. Examples of "Junk" Mr. Greenleaf lists are old paperbacks, rock records, cents off coupons, slide rules, McDonald's give-aways and old machine tools.
2. Collectible Phase: The wheel of

ELECTRIC REMOTE CONTROL BATTERY OPERATED ROBOT. The magic-marker store price of $1.98 still on the box reminds us that the right space toys have shown considerable increase in value. We paid $450 for this 25 year old copy of Linemar's extra special robot.

history turns and suddenly collectors rush to cart off as much as they can of what was last year's junk. Aggressive dealers jump on the bandwagon. If the collectible has real merits and sufficient "high price pieces necessary to attract wealthy and influential collectors and dealers" it may move on up the ladder if not "it may remain forever on the fringe." Examples are beer cans, steam toys, stock and bond certificates, postcards and old typewriters.

3. Advanced Trendy Phase: Books and articles on the subject appear and competing collectors quickly escalate the prices to surprising levels. As it is a new collectible no one is really certain what anything is worth but everyone feels prices will go even higher. Prices do continue to rise even for average or marginal pieces as well as the really important ones which are appearing with frequency due to the publicity and interest. Specialty dealers, speculators and trend setting auction houses are key players in this fast game. Many collectors of the item from the "Junk" phase either drop out because they can't pay the new prices or they sell out at a handsome profit. Examples of the "Trendy" Phase are: Schimmel carvings, primitive paintings, decoys, large pine tables, tin automobiles and 1960's sports cars.
4. Retrenchment Phase: The bloom is off the rose. When new collectors stop coming into the market and speculators get shakey about holding on in a market that raced up so fast and old collectors began selling the bottom of their collections the laws of supply and demand cause prices to fall. The best pieces decline the least, poor and average things the most. Collectibles in "Retrenchment" today are: most photographs, cast iron toys, 1950's automobiles, slot machines and advertising signs.
5. Mature Phase: Agreement has been reached by collectors and dealers about what is good and what is poor and prices stabilize. Speculators have no interest because prices do not skyrocket upwards. The best pieces may continue to rise faster than the inflation rate but average pieces are in much less demand. Small time collectors, dealers and pickers have a small role now because only a few pieces will come out of the woodwork. By now public awareness of the value of these items is so complete that when a piece does come to light a big time dealer or auction house will get it instead of the local flea market types. Examples of the "Mature Phase" are: Old Master prints, Federal furniture, Georgian silver, Tiffany lamps and automobiles made before 1930.

There is a danger that Mr. Greenleaf points out of putting too much emphasis on a small body of transactions thereby drawing conclusions that don't hold up over a period of time. Every antique and collectible has its regional rules as well. I know Ozark collectors who have just discovered cast iron toys, while East Coast collectors have viewed at least the more ordinary examples as being overvalued for several years. Robots are, for instance, already in an "Advanced Trendy Phase" on the East and West coasts and in Europe before many of the flea market dealers of St. Louis realize they are even saleable "junk". With the speed of information and travel working toward creating a global unity of all markets we may see many regional price differences

lessened in the near future. Price guides Eric Greenleaf mentions, play an important role in this even if the book is not well thought out. As collectors and dealers we are painfully aware of the responsibility of writing a book on a subject that is for the most part at the critical collectible phase. The flaws of this book are in part the flaws of suppositions based on too little evidence. Our remarks must be taken as mere directional indicators, not finished maps.

The "Product Life Cycle" is an important concept for collectors of space toys to understand. It governs the desirability, hence the price, of things. No one should collect anything for monetary reasons alone. All antiques and collectibles lack the liquidity of stocks, bonds, money market certificates and other financial instruments. Speculators can disrupt the orderly growth of an antique market as much as they can cause distress in a financial one by prematurely driving up prices, getting out and leaving everyone else with overvalued holdings. Over-enthusiastic and inexperienced collectors can cause unsupportable price rises themselves as anyone who attends auctions can attest to. Given today's volatile economic conditions one should not think that anything involving financial transactions, be it collecting space toys or growing apples, will be free from these contractions and expansions.

We do think you might profit from an understanding of what may lie ahead for the different categories of space toys. While the price guide will deal with those realities specifically we might generalize about how space toys fit into Greenleaf's five phases of collectibles and antiques. Playing the Devil's Advocate we raise objections to them at times not because we don't personally believe in space toys as legitimate collectibles but because we want the market to be at least somewhat objective. While as we said no one should view any collectible as a pure source of profit we know of few collectors who want to throw their money away. Paradoxically the collectors we know who could sell out at the greatest profit are apparently little concerned about how much something costs. What they care about is how good it is. But knowing what is good in an object is often not possible so early in the product cycle. Clearly most space toys are in most places in the second phase - "collectible." Prices are rising, information is coming out and yet it's still possible to get great buys from those who are a bit slower on the uptake. Space toys collectible phase items are: all plastic robots, recent robots, rockets, saucers, vehicles, nostalgia category stuff - i.e. Tom Corbett, Captain Video and space theme miscellany, things like lunch boxes and banks.

Even below that some space items may be in the "junk" phase. McDonald's recently put out a "Happy Meal" in a plastic flying saucer. Certainly this is at this time marginal as a collectible. Numerous new toys of rather generalized design are on a space theme - and they can be easily found at garage sales. Possibly the junk category may move up in time - if the advanced categories continue to progress toward full acceptance.

The best Japanese battery operated space toys, circa 1955 to 1975, mainly robots, are now in the "Advanced Trendy Phase" in cosmopolitan areas. At toy shows the very best pieces bring $1,000 or more, mint in the box. Speculators were the source of many of these good pieces but we suspect that speculators do not now hold many more of the finest such classic space toys in reserve. Speculators do have a lot of average robots and space toys (such as the

Chinese tin ones and 1970's Japanese models) but it seems unlikely that many speculators would resist turning $5 spent 10 years ago buying a Mr. Atomic from the dusty storeroom of a Poughkeepsie hardware store into $1,000 cash. As things heat up and as more collectors come into the market prices for medium and lower quality space toys will rise too fast and "Retrenchment" will inevitably come. Then the big question looms before us. Are space toys good enough to make it into the big time? Will they or any part of the category enter the blue chip area and go into the last "Mature" phase? There can be no doubt that some kinds of antique toys have already graduated to the top class. Auction prices of $10,000 to $20,000 may be trivial compared to the millions important Masters paintings fetch but it is necessary to realize that in the case of classic toys 5 figure prices are not being paid for one of a kind objects but rather for relatively inexpensive mass produced factory goods. High toy prices may prove to be an important indication of future acceptance of the esthetic worthiness of industrially manufactured objects in general. Certainly they pave the way for the acceptance of other kinds of toys. As the chapter on "Space Toys as Classics" indicates the best Japanese tin battery operated robots and astronautical toys have strong similarities to the premier pre-World War I tin German toys now bringing $5,000 to $20,000 at auction.

An even stronger indication of the possibility that space toys may become "Mature Phase" collectibles is the interest the artistic community shows in robots. Artists in the U.S. such as Edward Kienholz have constructed robot-like sculpture. In France Eduardo Paolozzi has extensively used the theme of mechanical beings. He has as well been a collector of robot material of all kinds and is active in encouraging books and research on the subject. In December, 1981 and January 1982, the Otis/Parsons Gallery of the Otis Art Institute of Parsons School of Design in Los Angeles exhibited robots from the collections of Larry Whiteley, Scott Singer and Stephen Sansweet within a toy exhibit curated by Kathryn Loye. This indicates an international intellectual and esthetic curiosity about the subject. Such elite attention often opens up popular acceptance and broad based support - necessary steps in securing permanent stature for a category of collecting interest. Perhaps the most serious criticism of space toys as important collectibles might be their recent vintage and origins as factory produced goods. But the trend is toward viewing things on their own merits and not according to their association with any supposedly virtuous past age or pre-industrial work method. Taken this way, rationally and openly, space toys may ultimately advance to be very respectable, mature collectibles. This book is not intended as the crowning summation but rather as a step in this evolution.

ROBOTS

Robots are today the hottest space collectible. Robots of all kinds seem to represent so many vital, timely issues and concerns that it would be possible to theorize that the toy ones share some of this attention. But many of the toys have their own charms and need little reference to anything else to be captivating. Put simply it is very easy to become a collector of the almost endless variety of little walking mechanical beings without thinking about the implications. If we do consider what real robots (i.e. computer directed mechanical beings) may represent in the years to come we may see the true dimensions of space toy collecting. The title of the first science fiction magazine expresses it — AMAZING.

MR. ATOMIC, CRAGSTAN, Japan, reported in 9" and 11" tall size, battery-operated, reported to have been made circa 1955/1960 but listed in 1962 Sears catalog. Made in two color versions, silver and blue with red trim. Mr. Atomic is considered one of the rarest robots and one of the most visually striking as well. It has great action with swinging arms, multi colored flashing lights, a beep-beep sound as it bumps and goes.

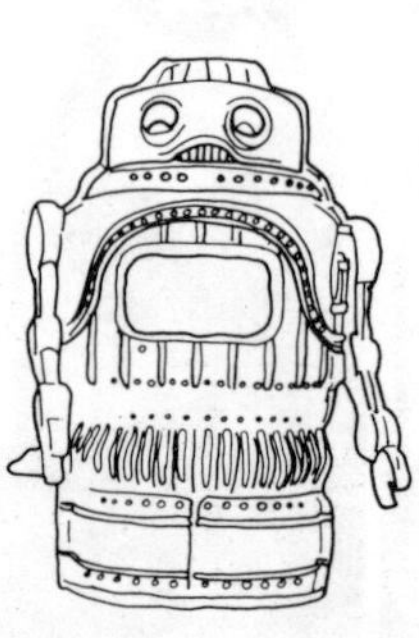

ROBOT, DAITO, Japan, 5" tall, wind-up, circa 1960/1965. A shy little robot that has an unusual design.

ANSWER GAME MACHINE, AMICO, Japan, 14½" tall, battery-operated, circa 1965/1970. This intellectual robot doesn't go anywhere. It can however help the kids with their arithmetic.

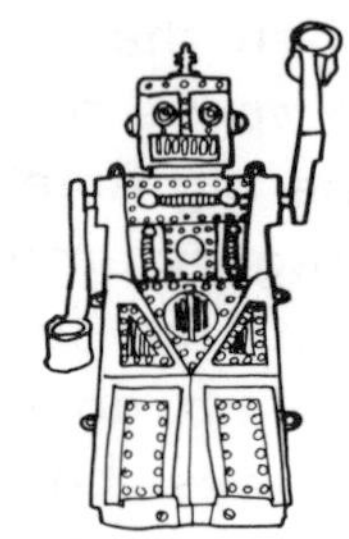

ROBERT THE WONDER TOY, IDEA TOY CORP., U.S.A. 14" tall, battery-operated, 1955. Robert was the toy of the giant robot star of the 1954 movie, TOBOR THE GREAT. It was then mass produced by Ideal in a remote control plastic version that talks, moves its arms and whose body is a tool chest.

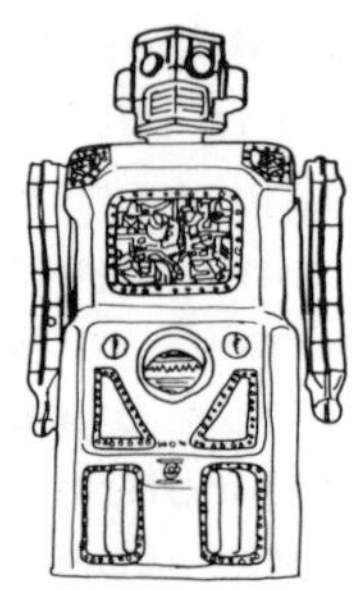

ROBOT, MT, Japan, 14" tall, battery-operated, circa 1960/1965. A Japanese version of Robert. The chest full of gears and cogs is lithographed.

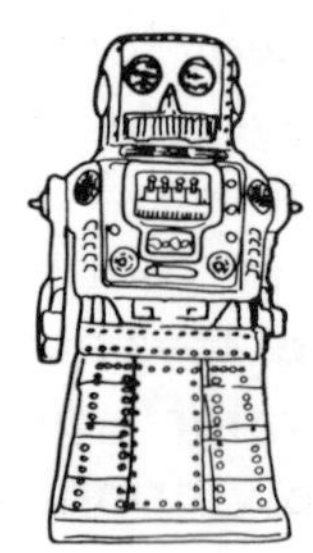

ROBOT, KO, Japan, 6½" tall, no action, circa 1960/1965. Yet another skirted "Robert" inspired Japanese robot.

LOST IN SPACE ROBOT, REMCO, U.S.A., 12" tall, battery-operated, circa 1965/1970. This is a licensed version of the robot from the TV series. Its hands operate by the controls in the back.

ROBOTANK-2, TN, Japan, 10" tall, battery-operated, circa 1960/1965. Robotank's arms pump back and forth then its two guns pop out and in a blaze of light they fire like machine guns.

MARS KING, S.H, Japan, 9½" tall, battery-operated, new in 1968 Horikawa toy catalog (S.H). The arms go up and we see on the TV in its chest a scene of Mars. The sound track is a "realistic" siren.

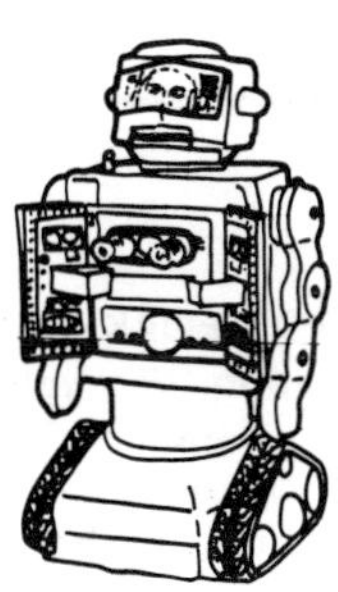

SPACE COMMANDER, S.H,-Japan, 10½" tall, battery-operated, new in 1968 Horikawa toy catalog (S.H). Another robot/man hybrid like the last but made a few years later. The face visor no longer swings open in this version, the chest door opens to the side and it is on a simulated tank tread base. As time goes on we often see more and more simplicity and loss of detail on robots probably to cut manufacturing costs. From the 1960's to 1980 the pay scale of Japanese workers has steadily risen, eventually pushing more and more toy making to Hong Kong, Taiwan and China.

ASTRO ROBOT, N, Japan,6" tall, wind-up, circa 1965/1970. Quite a realistic space suit of the American space program complete with flag shoulder patch and N.A.S.A. emblem.

ROBOT, Japan, 11" tall, battery-operated, circa 1960/1965. Somehow this boy's face behind the visor reminds one of the juvenile science fiction television programs of the early 1950's. At any rate the boy's personality is far less serious than the intrepid astronaut or the all work robots.

HI-BOUNCER MOON SCOUT, MARX, Japan, 11½" tall, remote control battery-operated, circa 1965/1970. Like Colonel Hap Hazard this very realistically suited astronaut has a revolving antenna on his head. HI-BOUNCER does something the real moon walkers didn't do - when the button on the remote control switch is pressed small plastic balls shoot out of a panel in his chest.

COLONEL HAP HAZARD, MARX, Japan, 11½" tall, battery-operated, circa 1965/1970. A very realistic astronaut in white space suit with red belt and black boots.

ASTRONAUT, CRAGSTAN, Japan, 13½" tall, battery-operated, circa 1955/1960. A big, well made toy with enormous presence. Without a doubt it is a classic. Comes in a red and a blue version.

ASTRONAUT, Japan, 8½" tall, battery-operated, circa 1960/1965. The batteries go in the tanks on the back that resemble scuba gear. This fanciful spaceman carries a TV camera, not a weapon.

MOON ASTRONAUT, DAIYA, Japan, 8½" tall, battery-operated, circa 1955/1960. A pre-moon walk toy design recognizable as such because of the deep sea diver-type helmet and science fiction-looking space suit in general.

ROBOT, S.H, Japan, 10" tall, battery-operated, circa 1960/1965. A red version of the above with slightly different detailing.

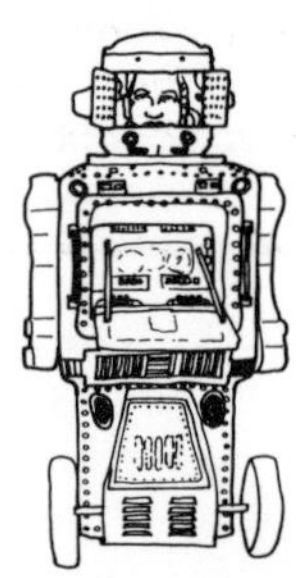

SPACE MAN, Japan, 6½" tall, battery-operated, circa 1960/1965. This small space man looks better equipped for hazardous police duty than outer space exploration.

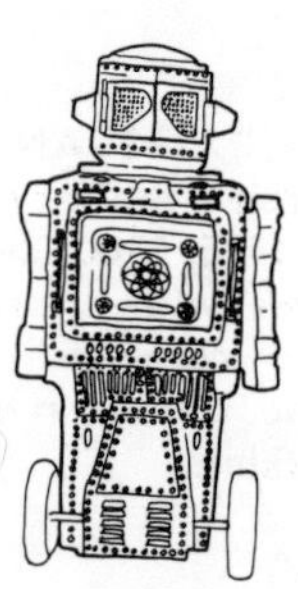

ROBOT, S.H, Japan, 10" tall, battery-operated, circa 1960/1965. This aggressive robot with a machine gun in the chest is also like the next revealed to be a man disguised as a machine when the visor pops open. This version is silver.

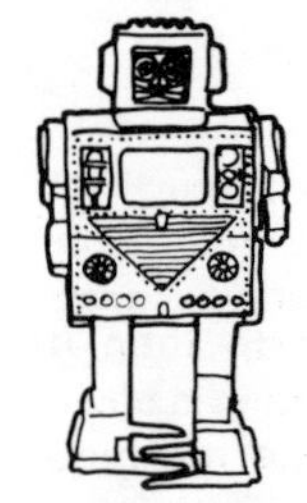

CAPTAIN ASTRO, Japan, 7" tall, wind-up, circa 1970/1975. Note the plastic feet on this small robot. It's a mechanism widely used today and is generally associated with lower priced, late robots.

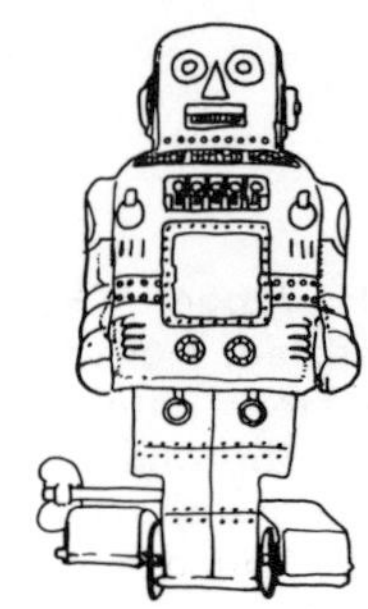

ROBOT, SY, Japan, 6½" tall, wind-up, circa 1960/1965. Small mechanical robots with "sparking" chests are not uncommon yet they are desirable if they are tin and have good lithography.

FIRE ROBOT, SY, Japan, 6½" tall, wind-up, circa 1960/1965. Is this a robot disguised as an astronaut or vice versa? At any rate he looks helpful and reliable as befits a fire fighter.

REMOTE CONTROL FORK LIFT ROBOT, S.H, Japan, 7" tall, battery-operated. New in 1976 Horikawa toy catalogue. A later all plastic robot (body yellow, fork lift red) that operates by remote control.

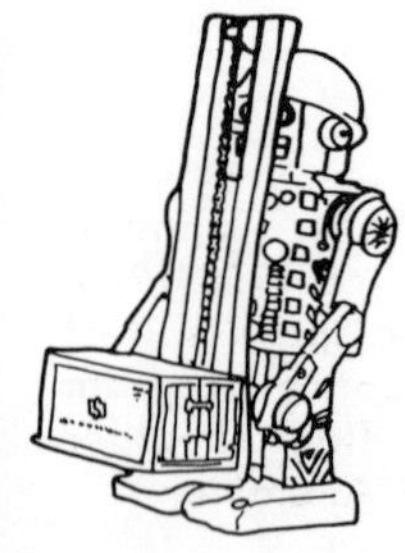

FORKLIFT ROBOT, S.H, Japan, 11" tall, battery-operated, circa 1965/1970. Very similar to Busy Robot with wheelbarrow except this hard hat robot operates a fork lift. Important and valuable, these early examples illustrate the trustworthy industrial robot concept.

BUSY ROBOT, S.H, Japan, 11" tall, battery-operated, circa 1965/1970. S.H has produced through the years several construction worker robots. The Busy Robot has a yellow hard hat, body and wheelbarrow.

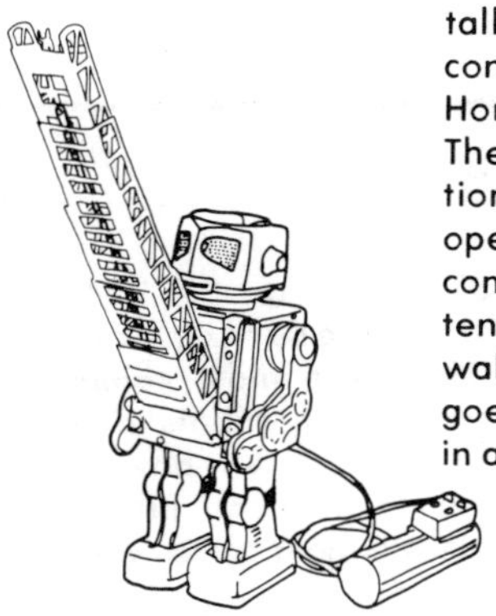

FIRE ROBOT, S.H, Japan, 11½" tall, battery-operated, remote control actions, new in 1971 Horikawa toy catalogue (S.H) The original catalogue description speaks for itself: "Walks, open/shut chest door, ladder comes out of chest and extends, then comes in. Also walks as ladder stands. Siren goes off." A most useful robot in an emergency.

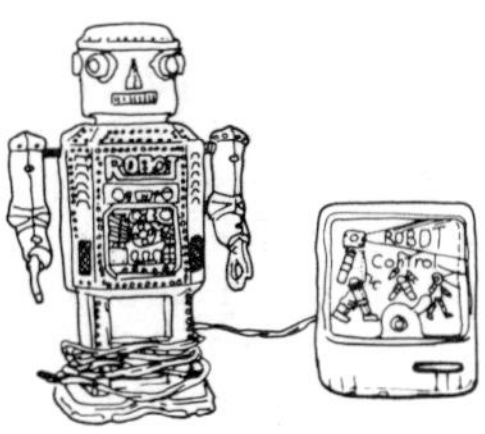

ROBOT, Japan, 7½" tall, remote control, battery-operated, circa 1955/1960. An early little robot with classic looks. Don't pass this one up at a garage sale - it's quite valuable.

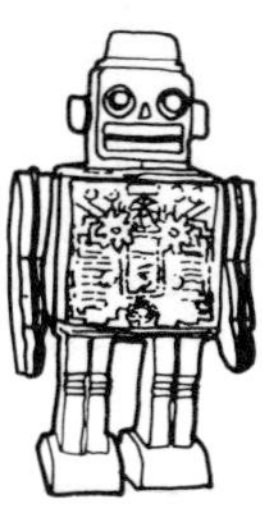

ENGINE ROBOT, S.H, Japan, 10" tall, battery-operated, circa 1960/1965. The Horikawa (S.H trademark) toy catalogue of 1968 lists this robot in a wind-up version (Art No. 936 Mech. Gear Robot) and a battery-operated version with slightly different see-through gear arrangement (Art. No. 916 - Engine Robot). No doubt a number of versions exist.

ROBOT, Japan, 10" tall, battery-operated, circa 1960/1965. Similar to the early S.H gear robots but with two antennas on the shoulders and a distinctive crest on the head.

BLING A GEAR, TAIYO, Japan, 15" tall, battery-operated, circa 1965/1970. The large size of this visible gear robot makes it a collector's prize.

HIGH WHEEL ROBOT, KO, Japan, 9" tall, battery-operated, circa 1960/1965. With blue tin body and red ears, hands and feet this is an interesting but not rare gear robot.

HIGH WHEEL ROBOT, KO, Japan, 9" tall, wind-up, circa 1960/1965. Very similar to the battery-operated version.

GEAR ROBOT, S.H, Japan, 9" tall, wind-up, listed new in Horikawa 1976 catalog (S.H). Again S.H makes a robot with visible gears in both wind-up and battery versions. The dominant material is plastic as compared with the late '60's tin ones.

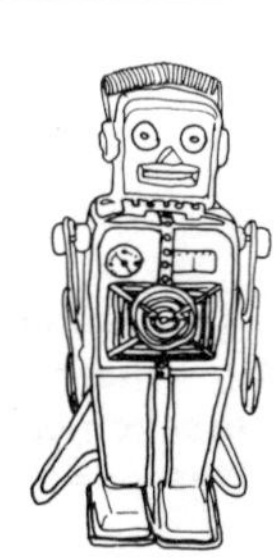

ROBOT, Y, Japan, 6" tall, wind-up, circa 1955/1960. An amusing, mechanical man with lively actions.

SUPER SPACE GIANT, S.H, Japan, 16½" tall, battery-operated, circa 1965/1970. An almost identical version appears in the 1971 Horikawa toy catalogue (S.H) as the Super Space Giant so likely that was the very descriptive name for this very large robot that offers, according to the catalogue, the following features - "Stop'n go, swing open door, blinking and shooting gun, realistic shooting noise, rotating body." It came, incidentally, 6 to a carton, not the usual dozen or two.

ROBOT, RFA, 7½" tall, wind-up, circa 1955/1960. An early, very machine-like mechanical robot.

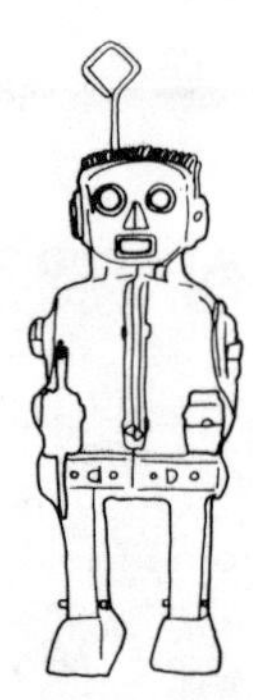

ROBOT, Japan, 7½" tall, wind-up, circa 1955/1960. A less common variation of the primitive robot this one comes with wrench ready to work. All these simple early silver robots have a 1930s pulp magazine cover flavor. Sparky's box is a dead giveaway that the toy designers of Japan were well informed about science fiction pulp and pop films of the 1930s and late 1940s.

ROBOT, Japan, 5" tall, wind-up, circa 1955/1960. Another variation of the mechanical robot that sparks. This one is blue.

SPARKY ROBOT, KO, Japan, 6½" tall, wind-up, circa 1955/1960. The box shows a hundred foot robot lumbering down a city street derailing street cars, shining his eyes like search lights on the night sky behind him. When we open the box we find an altogether less formidable robot. Sparky's antenna turns him on when raised.

ROBOT, Japan, 8½" tall, battery-operated, circa 1960/1965. The most common arm action in toy robots is from side to side. A few however, like this model, can lift their arms. It is almost a "so what?" expression.

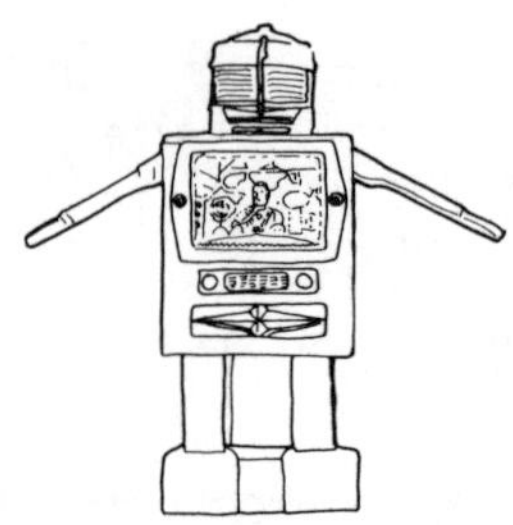

SPACE EXPLORER, Y, Japan, battery-operated, circa 1960/1965. At the controls in the chest cavity of this robot is an astronaut. It is an unusual piece. The robot he's driving even has a vague resemblance to an early 1960's Detroit automobile.

ROBOT, Japan, 6½" tall, wind-up, circa 1965/1970. A delightful little mechanical man whose arms go up and down.

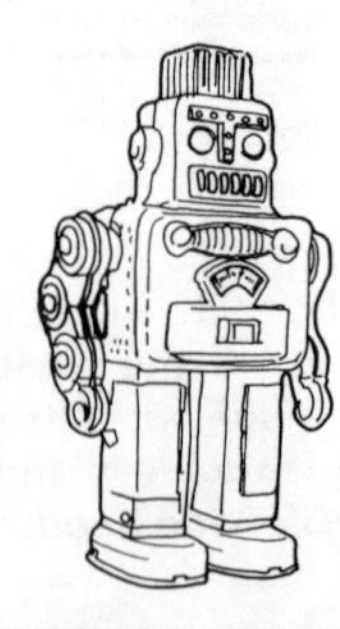

SMOKING SPACEMEN, LINEMAR, Japan 11½" tall, battery-operated, circa 1955/1960. As the box reports it has "lighted eyes, stop and go action, lighted pistons" and best of all smoke issues from his mouth. In short, a classic toy.

ROBOT, M, Japan, 11" tall, battery-operated, circa 1965/1970. A wonderfully restrained robot in silver and red. Very elegant - and valuable.

LUNAR ROBOT, Y, Japan, 7" tall, wind-up, circa 1965/1970. A small mechanical robot that appears to be smiling. The face is mounted directly on the legs with what might be normally an upper body above the head.

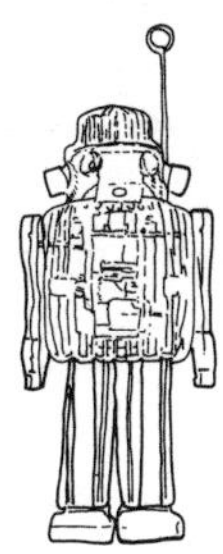

ROBOT, Japan, 8" tall, battery-operated, circa 1965/1970. With red legs, silver arms and yellow ears this unmarked robot has a transparent chest and head revealing its internal mechanism.

PLANET ROBOT, KO, Japan 8½" tall, battery-operated, circa 1955/1960. A version of Robby the good robot of FORBIDDEN PLANET and THE INVISIBLE BOY. All the early tin Robbie's are desirable.

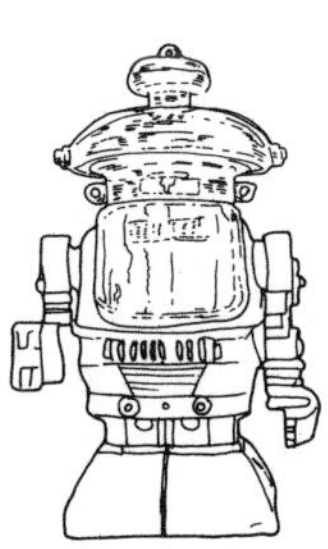

ROBOT, Japan, 8½" tall, wind-up, circa 1965/1970. A cousin perhaps to the LOST IN SPACE television robot.

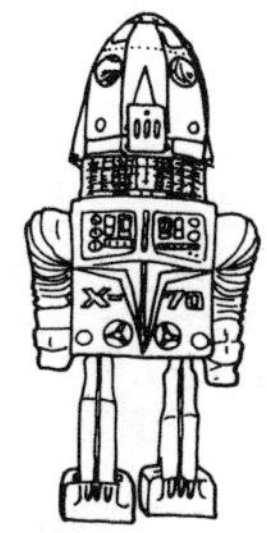

ROBOT X70, TN, Japan, 12½" tall, battery-operated, circa 1960/1970. A robot with a pop open head, orange arms, and red feet. Not a standard design but quite striking.

THUNDER ROBOT, ASAKUSA, Japan, 11" tall, battery-operated, circa 1960/1965. This strange mechanical being seems more appropriate to the medieval fantasy world of Hieronymus Bosch than the toy boxes of the 1960's.

DINO-ROBOT, S.H, Japan, 11" tall, battery-operated, circa 1965/1970. Dino is a reptilian monster disguised as a robot. An unusual concept and a desirable collectible.

SIR GALAXY, MATTEL, U.S.A., battery-operated, wireless radio-controlled, 1978. Originally designed by International Topper Corp. Mattel purchased the rights and brought out a version with minor modifications. It can walk, turn and transmit the apparatus' voice. In white and black plastice.

MIGHTY ROBOT, N, Japan, 5" tall, wind-up, circa 1965/1970. This small wind-up astronaut with the spinning device on his head is much rarer than similar ones that lack that feature.

DRIVING ROBOT, SY, Japan, 6½" tall, wind-up, circa 1960/1965. Another intemperate robot with a slightly maniacal expression - as though somewhere a nut is loose.

ROBOT, Japan, 6½" tall, wind-up, circa 1960/1965. What amuses a robot? (unanswered). What does the skull of a robot look like? (answered: the central column of this piece pictures a robot with gaping eye sockets). Mixing the diabolical and the manic we have a wonderful toy almost medieval in flavor.

MEGO MAN, SY, Japan, 6½" tall, wind-up, circa 1955/1960. A sublimely happy robot that rings a bell as it rolls along. With his red face and half closed eyes it looks as if he's been drinking - what compound can put a buzz on a robot?

ROBOT, YONE, Japan, 3" high, wind-up, circa 1970/1975. This tiny, droll robot wouldn't look out of place under a toad stool in a fairy tale. As science fiction took a Tolkien-esque turn little mythological robots mysteriously began appearing.

ROBOT DOG, KO, Japan, 6" long, wind-up, circa 1955/1960. What robot wouldn't like its very own pet robot dog. What robot collector wouldn't take good care of this little metallic mongrel for that matter.

SPACECRAFT

Rockets that are used to escape the gravitational pull of the earth must of necessity be streamlined. The essential traditional rocket is in fact a shape that offers little wind resistance. Once in space however there is no friction from air so any design will do even one that looks as aerodynamically awkward as the Apollo Lunar Module. Past science fiction visualizations of sleek swept-wing spaceships in fact contrasted markedly with the reality of the nuts and bolts spacecraft of the 1960's and 1970's.

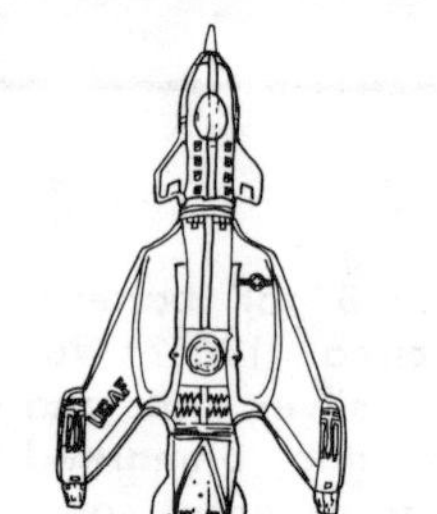

UNIVERSE ROCKET, TN, Japan, 21½" tall, battery-operated, circa 1960/1965. An impressive rocket rather like artists' conceptions during the late 1930's. The red and black trim on the white body complements the streamlined shape.

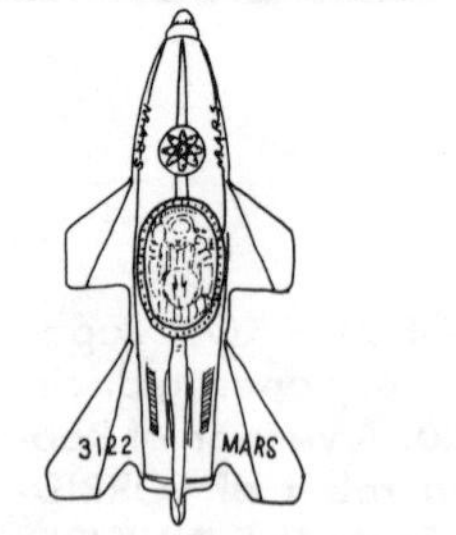

MARS ROCKET, MT, Japan, 14" tall, battery-operated, circa 1960/1965. Very similar to the 1944 German A-4b which was a winged V-2 capable of long glides. Wings on rockets were to diminish in fact until the space shuttle which of course needs to be able to glide within the atmosphere of Earth.

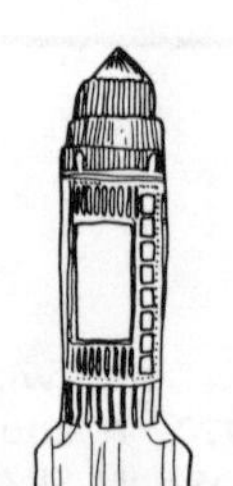

ROCKET, Japan, 11" tall, battery-operated, circa 1965/1970. A toy rocket somewhat restrained yet lacking in realism and specificity at the same time. Too generalized to attract much collector interest.

ROCKET FLIM, Hungary, 15½" tall, friction, circa 1960/1965. If the rocket runs into something it stands up on its tail fins. Space toys were made in the Soviet Union and in Soviet bloc countries but the authors were unable to learn much about them. An almost identical friction rocket was made in Japan by SM.

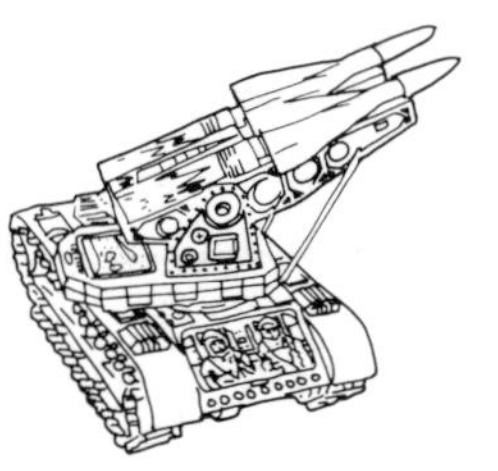

SPACE GUARD, MT, Japan, 9" long, battery-operated, circa 1965/1970. Two astronauts operating two very lethal-looking missiles atop a tank. The history of the rocket is inseparable from its use as a weapon as this toy reminds us.

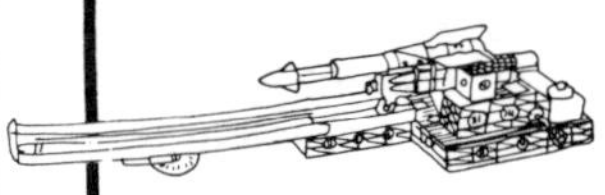

LAUNCHING PAD, Japan, 19", spring-operated, circa 1960/1965. Except for the bright blue, yellow and red decoration this is a rather military looking affair.

EARTH SATELLITE, MAR, Japan, 9½" long, circa 1960/1965. It's easier to paint a name on a toy than make it look like the real thing.

ATOMIC ROCKET, MT, Japan, 7" long, friction, circa 1955/1960. A hold-over from the Buck Rogers era. Japan made many small tin rockets after the war that lack specific identity but have well done trim and designs.

FIRE BIRD, Japan, 13½" long, battery-operated, circa 1960/1965. A number of strange boat, car, rocket fantasy vehicles were made in the 1960's. Some like this one are even controlled by the sound of a big plastic whistle which came with them.

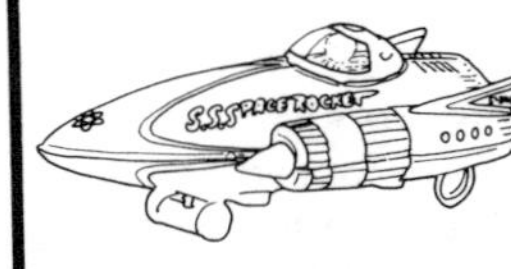

S S SPACE ROCKET, KO, Japan, 13½" long, battery-operated, circa 1965/1970. Is it a plane, a boat, a car or a space rocket?

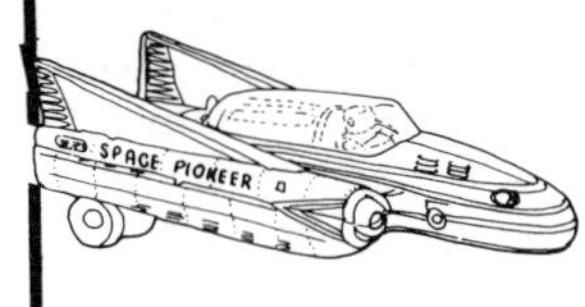

SPACE PIONEER, MT, Japan, 12" long, battery-operated, circa 1965/1970. More a futuristic sports car than a spaceship this streamlined vehicle ignores the essential fact that real spaceships are so expensive they must of necessity be the property of countries, not individuals. The Buck Rogers myth of private travel in space dies hard when fantasies as slick as this are produced.

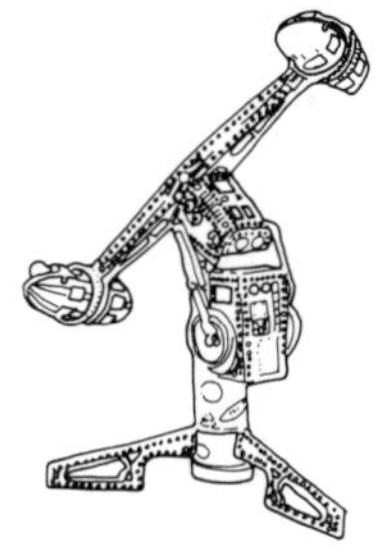

ROCKET RIDE, ALPS, Japan, 15" tall, battery-operated, circa 1960/1965. Related more to a carnival ride than the space program. A nice toy with such good action its irrelevance to space or science fiction must be overlooked.

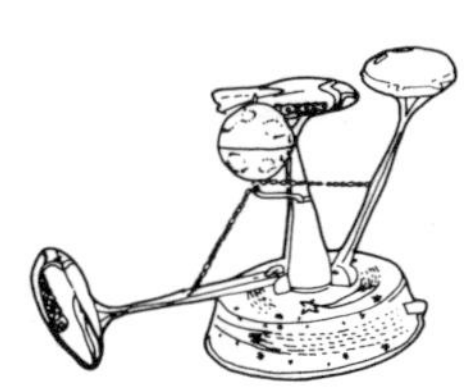

CAROUSEL, JOUSTRA, France, 6½" tall, wind-up, circa 1955/1960. Spacecraft circle a tin moon in this toy that vaguely resembles an orrery - the 18th century device that demonstrated the moon and earth moving in orbit around the sun.

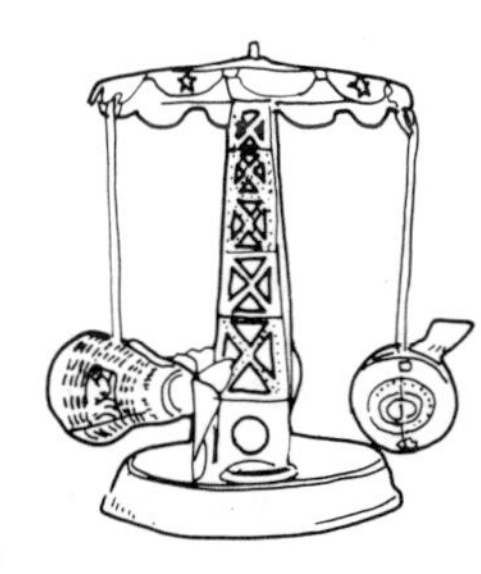

CAROUSEL, Japan, 7½" tall, windup, circa 1965/1970. A traditional carnival ride-type toy with space capsules instead of horses or airplanes.

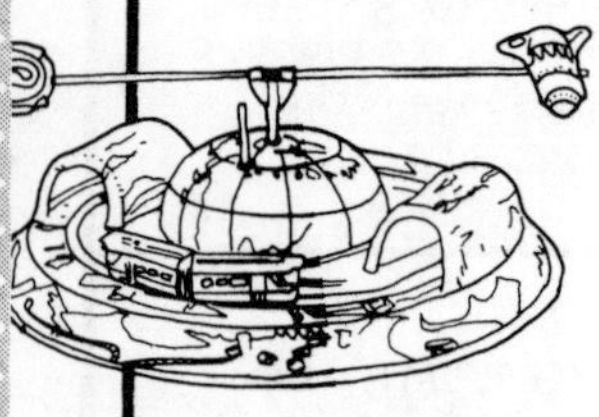

CAROUSEL, Greece, 10" across, wind-up, circa 1960/1965. As a rule Greek and Turkish tin toys are primitive compared to the Japanese. This little toy with a train and a rocket circling around a globe may have a certain poetic charm but would pale beside the classic 1960's Japanese space toys.

MAN MADE SATELLITE, SKK, Japan, battery-operated, circa 1958. Obviously an attempt to use the newsworthy subject of satellites. The early Sputniks and Vanguards had projecting antennae like this. Scientifically important though they are the orbiting radio transmitters proved to be ditticult to translate into exciting toys.

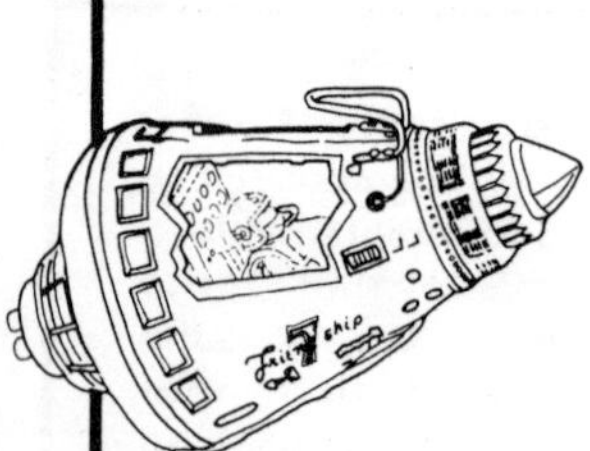

FRIENDSHIP 7, S.H., Japan, 9½" long, friction, circa 1962. The first American to go into orbit around the earth was John Glenn, launched on February 20, 1962. The modified Atlas IBM was Mercury 6 and the space capsule was named Friendship 7. The truncated cone shape is rather nicely decorated. An interesting feature is the windows that look like the toy designer interpreted a cut-away diagram to be real windows. The best feature is that when the toy is pushed along and the wheels turn the little tin astronaut rotates inside in a simulation of weightlessness.

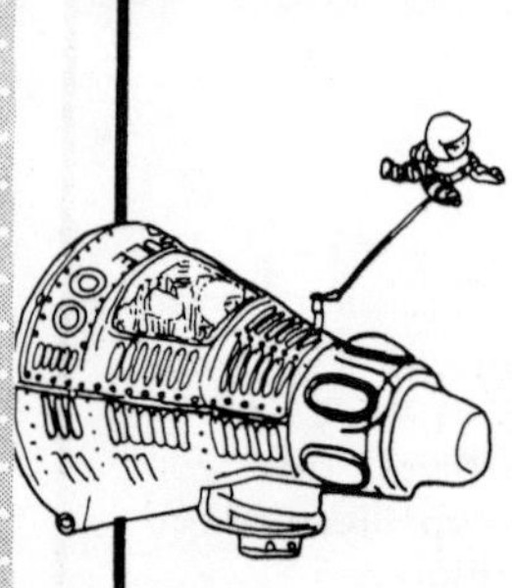

CAPSULE 7, MT, Japan, 9½" long, battery-operated, circa 1965/1970. Both the Mercury and then the later Gemini spacecraft looked essentially alike — basically a truncated cone. A few liberties with the design have been taken with this toy to good effect. With the plastic astronaut taking a spacewalk it is quite a worthwhile piece.

CAPSULE, SM, Japan, 6" long, friction, circa 1960/1965. A simple but effective small friction capsule with a gaudy yellow and red suited astronaut dressed more like a circus lion tamer.

CAPSULE, SH, Japan, 6" long, friction, circa 1960/1965. Two astronauts forever orbit in this little toy Mercury-inspired spacecraft.

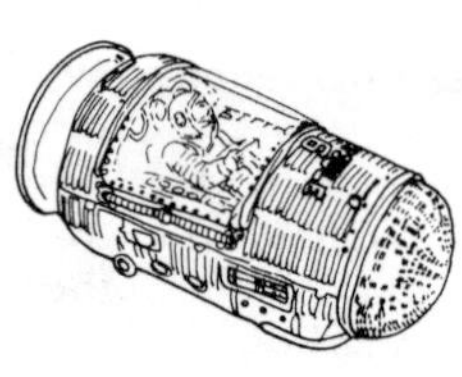

CAPSULE 6, MT, Japan, 11½" long, battery-operated, circa 1960/1965. More authentic in its details than its shape this astronautical toy still has a decidedly real spaceship feeling. Inauthentic, of course, is the astronaut's 35mm camera — the official space camera was the 2¼ X 2¼ Hasselblad. This perhaps takes the Japanese love of the miniature camera to new heights.

EAGLE, DSK, Japan, 10" tall, battery-operated, circa 1969. Between 1969 and 1972 12 American astronauts walked on the moon. They landed on the moon in a Lunar Module spacecraft after being launched by a Saturn V rocket and travelling in an Apollo Command and Service module (the Command module also served as a re-entry craft). "Eagle" was the name of the Lunar Module on the Apollo 11 mission to the moon. In it Neil Armstrong and Edwin Aldrich landed on the moon on July 20, 1969. When Armstrong descended the Eagle's ladder and jumped down he became the first human in history to set foot on the moon. This toy is quite a good representation of the actual spacecraft.

SPACE STATION, TN, Japan, 11½" long, battery-operated, circa 1960/1965. While not a replica of any of the orbiting scientific laboratories called space stations this toy has a generalized look of both the USSR's Salyut Space Station and the US' Skylab.

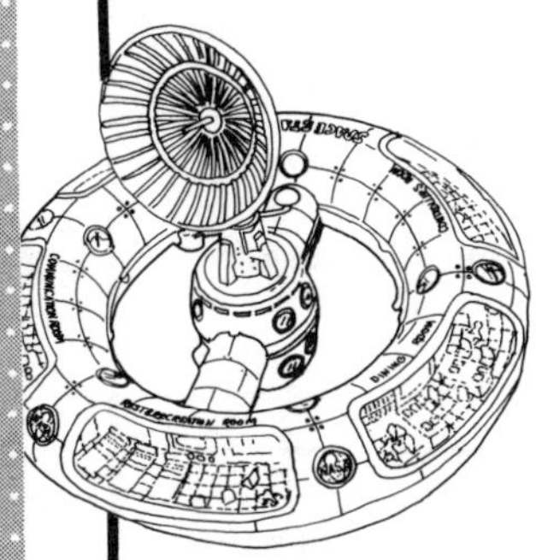

SPACE STATION, S.H., Japan, 11" wide, battery-operated, circa 1960/1965. Various donut-shaped space stations were envisioned by space scientists from 1928 on. Although the space stations to date have been straight cylinders it may yet come about that great wheel-shaped manned satellites come to orbit the earth. This is an imaginary design but not a totally fanciful one.

SAUCERS

Toy flying saucers come in a variety of sizes and colors but they all have one thing in common — they resemble no actual space craft. This being so, fantasy has no limits in saucer design. Yet as they have little or no tie-ins with comic or movie characters and few of the philosophical overtones of robots collectors do not place a premium value on them today. Even the early ones are usually priced below $200. Still we find them a fascinating study in fantasy toy design and many are quite beautiful as they bump and go in a pulsing blaze of flashing colored lights.

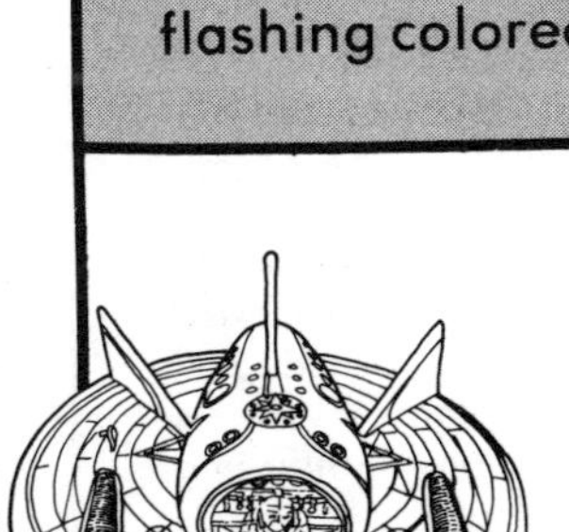

SPACE GIANT, MT, Japan, 12" long, battery-operated, circa 1960/1965. Lots of early 1960's Japanese toys had tail fins as prominent as 1950's Detroit automobiles.

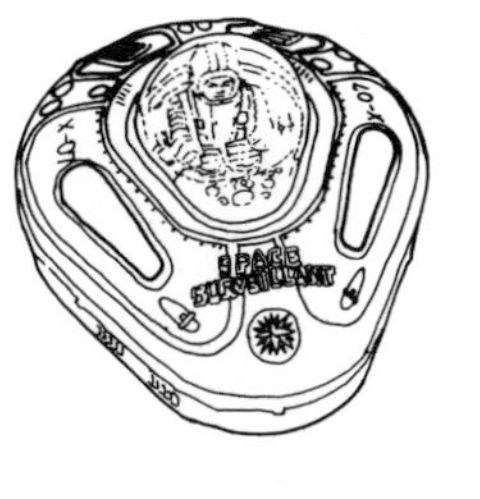

SPACE SURVIVOR, MT, Japan, 8½" long, battery-operated, circa 1960/1965. With a slightly streamlined shape this saucer is marked X-07 like an experimental Air Force test plane and has a compass-like front ornament worthy of a deluxe early sixties Chrysler.

SAUCER, NeKur, Turkey, 6½" long, friction, circa 1960/1965. Made in Turkey this toy is distinctly inferior in design and construction to the better Japanese toys of this era.

SAUCER, Japan, 7" long, friction, circa 1955/1960. An amusing little toy that looks more like a surfboard than a flying saucer.

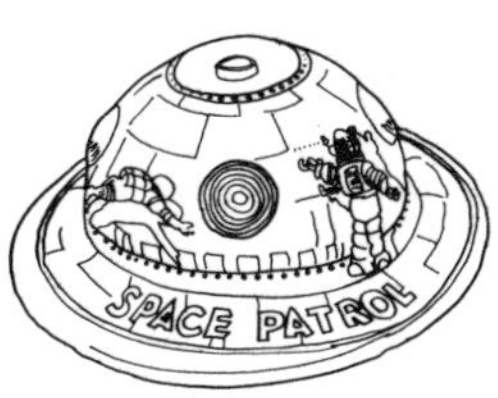

SPACE PATROL, MT, Japan, 8" long, battery-operated, circa 1960/1965. With a Robby-like robot dancing on the side and a deep sea diver-type space man floating beside him this is a typically eclectic saucer. Space Patrol of course has no reference to the TV show. This has the air feeder on the top which supported a ball or paper figure on a cone of air when operated.

EXPLORER SHIP, MT, Japan 10½" long, battery-operated, circa 1965/1970. MT (K.K. Masutoku Toys of Tokyo) has made a saucer with 6 multi colored lights surrounding it for years changing the uniform of the little pilot (or pilots) to keep up with the times.

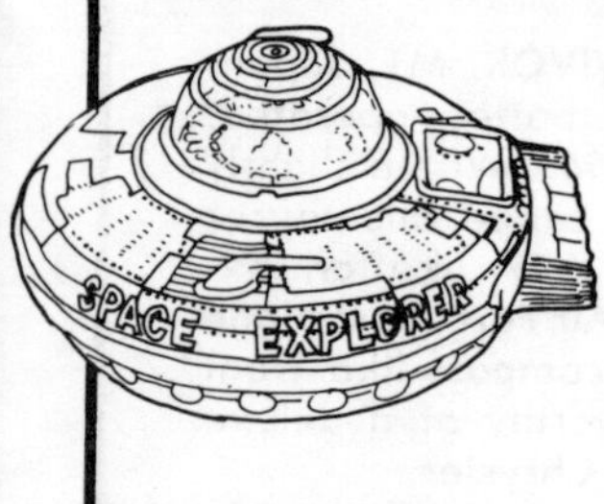

SPACE EXPLORER, ICHIKO, Japan, 8″ long, battery-operated, circa 1965/1970. Saucers provide a large area for lithography and few are alike in the details of rivets, panels and insignia.

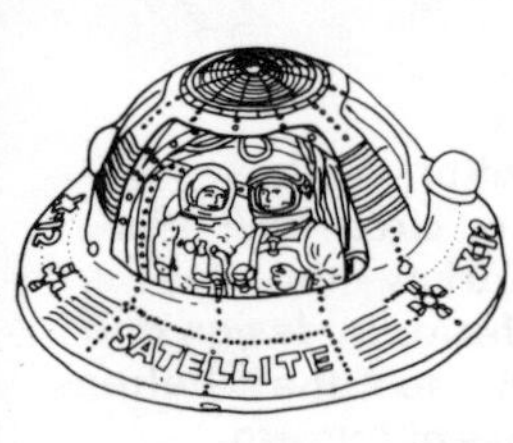

SPACE SURVEYOR X12, MT, Japan, 8″ long, battery-operated, circa 1960/1965. Yet another flying saucer taken from the hands of little green men and pressed into service by astronauts to explore space.

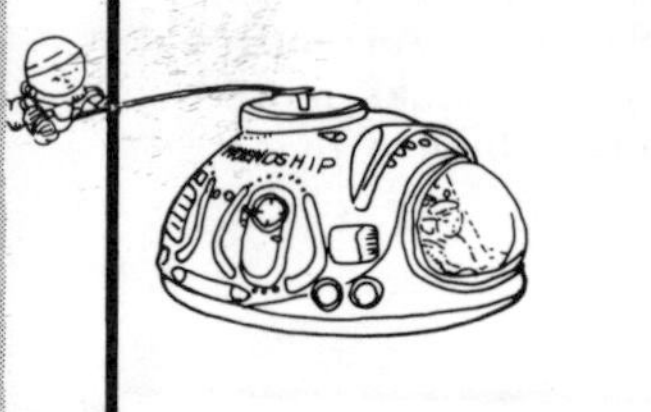

FRIENDSHIP, Japan, 8″ long, battery-operated, circa 1960/1965. No doubt referring to the Mercury capsule called "Friendship" that carried Lt. Col. John Glenn in 1962 as he became the first American to orbit the earth in space, this Friendship bears no resemblance to the actual capsules but does have a certain charm of its own.

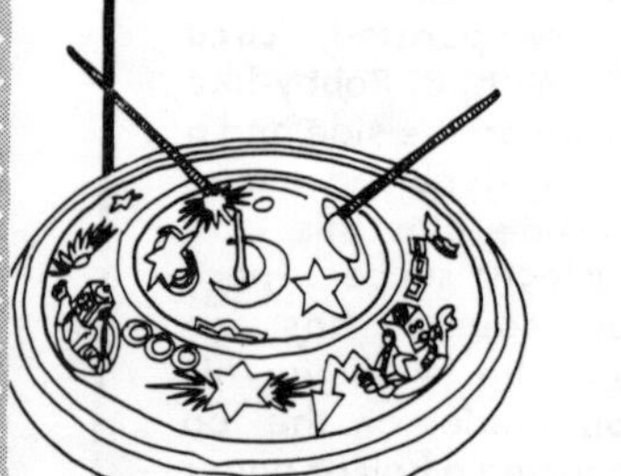

SAUCER, RFA, 4″ long, friction, circa 1955/1960. An early saucer of very hokey design. Such generalizations are not looked upon with favor by most collectors.

VEHICLES

The space theme collides with more traditional toys in several of these. Some could be in other categories and several are very marginal indeed to any sort of space toy collection. We include them to illustrate some forms of space toys that may not have much future.

CAPE CANAVERAL, CRAGSTAN, Japan, 8½″ long, battery-operated, circa 1955/1960. Although a conventional form the specificity of the lettering gives this some credibility as a space collectible.

MARTIAN SUPER SENSITIVE, Japan, 9½″ long, battery-operated, circa 1955/1960. This mixture of a jeep, a strange radar apparatus, and a space man cannot help but amuse us. A nice if strange adaptation of the space theme.

SPACE PATROL, MT, Japan, 9½″ long, battery-operated, circa 1960/1965. The ubiquitous words "Space Patrol" are boldly placed on this rather odd combination of a 60's Ford-look vehicle driven by a deep sea diver/space man with a lethal looking machine gun mounted on the hood. Hopefully Mr. Space Patrol won't cut in front of the red robot-driven Martian Buick. A fire fight between these types would leave the freeway closed for days.

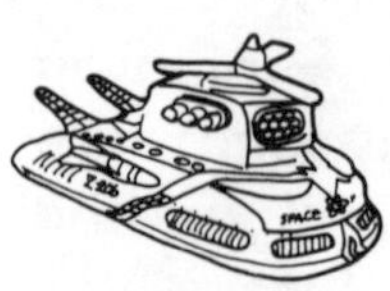

FLASH PATROL, TPS, Japan, 7½″ long, battery-operated, circa 1965/1970. Is it a helicopter, a car, a boat, a space ship? Not an uncommon relatively new toy of uncertain purpose.

SUPER CYCLE, B, Japan, 12″ long, friction, circa 1955/1960. "Space Patrol" like the atomic symbol was affixed to Japanese space toys wherever there was a space not occupied by rivets or dials or gauges. This futuristic motorcycle is driven by an all green man in a helmet.

SPACE PATROL, Japan, 12" long, battery operated, circa 1965/1970. A spaceman driving an atomic powered VW Beetle! What a droll toy. Perhaps the spaceman is a collector of vintage automobiles as were several of the test pilots of stratospheric experimental aircraft some years ago.

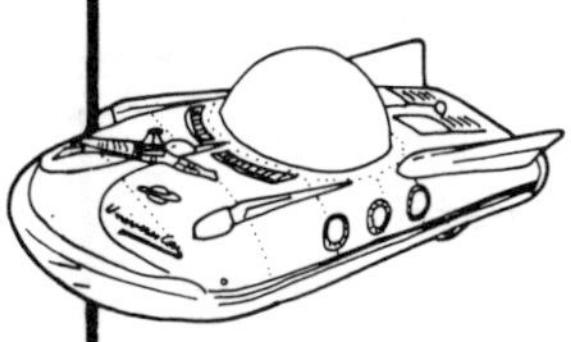

UNIVERSE CAR, China, 10" long, battery operated, circa 1970/1975. In the 1970's a number of tin toys came from China that resemble the Japanese toys of the 1950's and 1960's. It may be too soon to know what place these will occupy in collectors' hearts and cabinets.

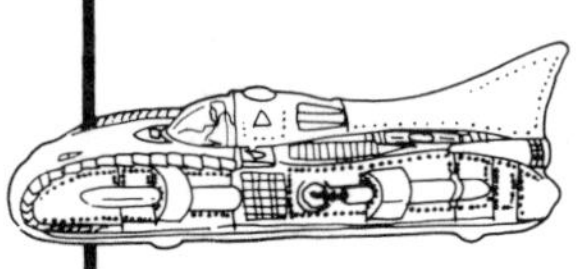

HOLDAUTO, Hungary, 12" long, battery-operated, circa 1960/1965. Several space toys blow a circle of air that can suspend a paper figure or a ball in the air as they travel along. This is one such toy - in the form of a futuristic car.

SPACE EXPLORER, TN, Japan 12½" long, friction, circa 1960/1965. This is little more than a revised diesel train engine. Such modifications of existing toys usually are not as desirable as designs that are completely original to the theme.

BUS, China, 7½" long, friction, circa 1970/1975. A toy with marginal appeal as a space item. Not highly valued by collectors, especially if they lack charm.

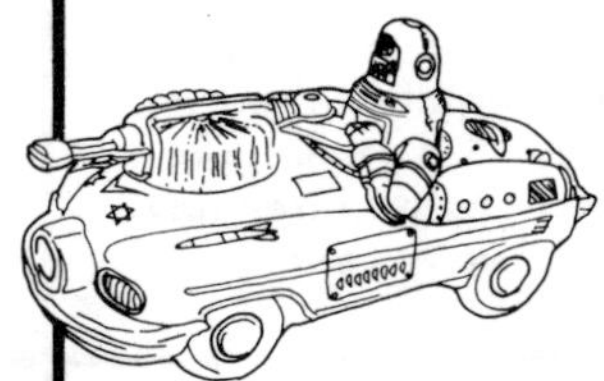

AUTOMOBILE, TN, Japan, 8" long, battery-operated, circa 1955/1960. A red robot in a Martian version of a 1950's blue Buick? This particular robot doesn't look too tranquil so I'd pull over if he gets behind you on the canal freeway.

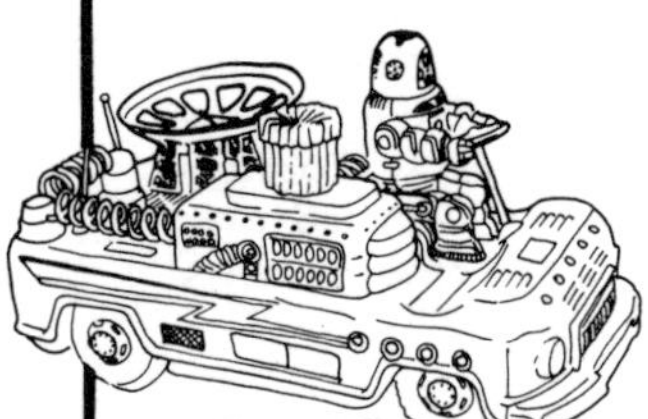

SPACE ROBOT, Y, Japan 9½" long, battery-operated, circa 1955/1960. Several versions exist of this wonderful vehicle full of radar dishes, coiled wire and light up plastic domes. The competent robot driver will no doubt safely deliver this paraphernalia anywhere.

SPACE TANK ME-091, China, 9½" long, battery operated, circa 1965/1970. Although without a weapon a tank is a military vehicle by definition. This robot-driven example produced in China is abundant and not highly valued at present.

SPACE TANK, KO, Japan, 6" long, battery operated, circa 1955/1960. Very similar to the Chinese Space Tank but valued at about 5 times as much by collectors. Chinese space toys are frequently available new in the box at reasonable prices. They often closely resemble tin Japanese toys of a generation before.

SPACE PATROL, MT, Japan, 8" long, battery operated, circa 1960/1965. Several versions exist of this space tank. The lithos of rockets leave no doubt that this is a serious military weapon.

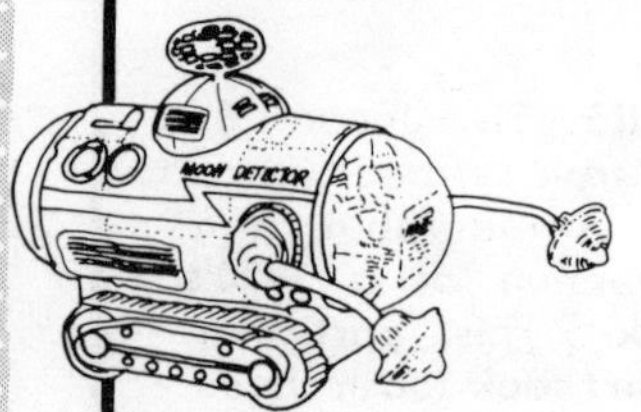

MOON DETECTOR, Y, Japan, 11" long, battery-operated, circa 1965/1970. An imaginative vehicle on caterpillar tread inspired by N.A.S.A. lunar program but not copied after any moon crawler we've seen.

ROBBY SPACE PATROL, TN, Japan, 12½" long, battery-operated, circa 1955/1960, Possibly one of the rarest space toys. The design resembles the landcraft Robby operated in the classic science fiction movie FORBIDDEN PLANET. Possibly this was even a licensed product - most of the other Robby's are not.

MOON PATROL, TN, Japan, 12½" long, battery-operated, circa 1955/1960. Besides this MOON PATROL SPACE DIVISION NO. 3 there is a MOON SPACE SHIP with a transparent ball atop the column as well as the very desirable ROBBY SPACE PATROL landcraft.

SPACE WEAPONS

Although the role models for the space explorer (Buck and Flash) were never without their ray guns the exploration of space has been (at least on the surface) free from weapons and aggression. Space toys continue to be more derived from fiction than fact so guns of all kinds pop up (literally at times) in the genre. Some of the robots and astronauts carry weapons as well.

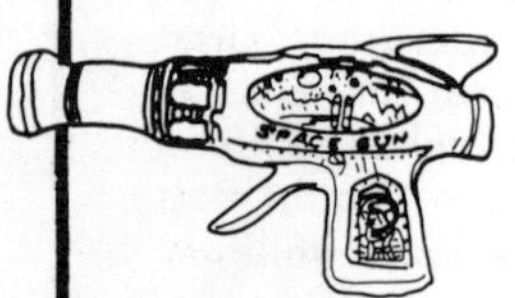

SPACE GUN, DAIYA, Japan, 7½" long, friction-drive with sparking mechanism, circa 1960/1965. A naive painting of the moon, a tail fin, a rocket shaped barrel — how could a kid not vaporize any hostile aliens hiding under his bed?

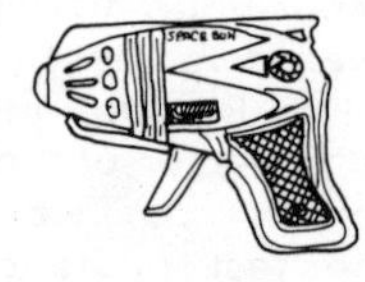

SPACE GUN, SHUDO, Japan, 4" long, friction-drive with sparking mechanism, circa 1965/1970. Moderately spacey. Numerous versions of these friction-drive space pistols exist.

SPACE GUN, SHUDO, Japan, 5½" long, friction-drive with sparking mechanism, circa 1970/1975. A rather more subdued space gun — more of a futuristic automatic. An identical version was produced recently in Korea by MTV.

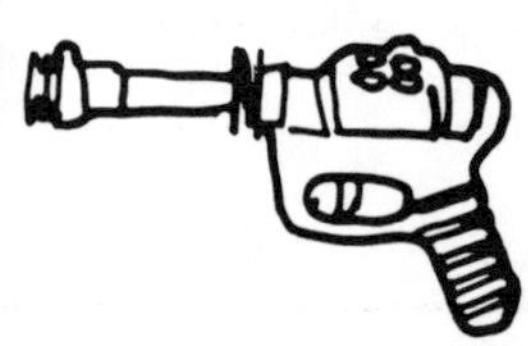

Y2-38 DISINTEGRATOR PISTOL, DAISY, U.S. Sparking mechanism. First made in 1935 this is the original Buck Rogers "Zap" gun.

Y2-44 LIQUID HELIUM WATER PISTOL, DAISY, U.S. Made first in 1936 this BUCK ROGERS IN THE 25TH CENTURY licensed beauty first came out in red and yellow (above). Later versions have a copper finish (below). It fired 15 times from its leather water pouch. The BUCK ROGERS Daisy guns were a terrific success — even in the Depression they sold millions. Today examples in mint shape are tough to come by.

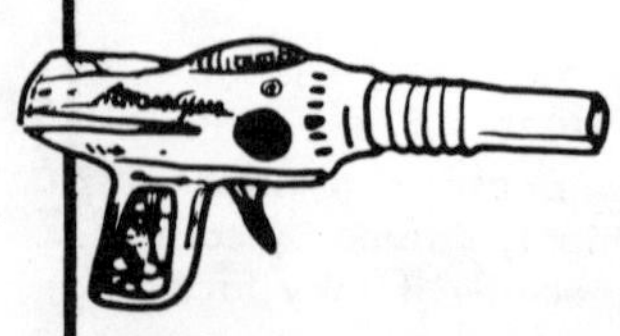

ATOM GUN, KO, Japan, sparking mechanism, 1965. A gun with a cockpit on top, atomic symbols on the side and an outer space picture on the grips. These imaginative little guns are still abundant and yet have a lot of charm as fantasy designs.

INTERVIEW WITH LARRY WHITELEY OF WHITELEY-LOYE GALLERY, LOS ANGELES, CALIFORNIA

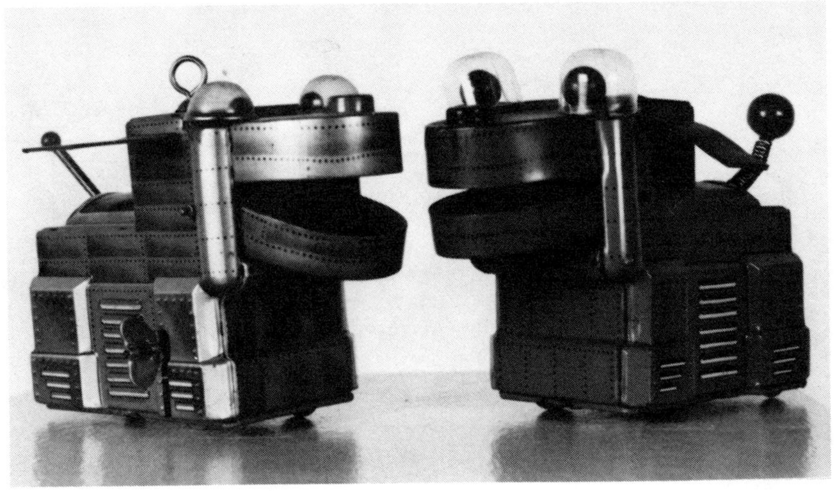

ROBOT DOG, KO, Japan, 4″

WHAT ATTRACTED YOU INITIALLY TO COLLECTING SPACE TOYS?
Initially I collected really early, 1850 to 1900, American toys and some German toys. I had a friend who collected the Japanese toys from the '30's and '20's and '40's and I thought that was really trashy looking stuff. But when I saw a whole wall of it one time it knocked me out. It was this beautiful color wall paper almost — instead of the fact that they were actual toys. The color of this whole wall of this yellow, the Japanese blue and red — all those colors together that they typically use on all the Japanese tin lithos. It just knocked my socks off. Then one day I was in a store where they sell out lots of stuff from gift stores and toy stores and I saw a real colorful robot. I saw him as a piece of sculpture, not just a toy, so I bought the whole case and boy it was all downhill from there. And interestingly enough for 4 years I was never telling anybody that I was collecting those things because it was like you were in the closet — you were hiding this stuff. Not that I didn't want anybody to start collecting 'em but I was embarrassed about collecting them, to tell you the truth. But I still saw them as beautiful objects. Because I had this reputation of selling and buying and owning in my own collection just some of the best toys that ever existed but it's funny that independently about 4 or 5 other of the top toy collectors in the country were also collecting robots under the table. And we all sort of came out of the closet about 3 years ago, or 2 1/2 years ago, and now the thing is exploding. . . .

WERE YOU A SCIENCE FICTION FAN AS A KID?
No, I wasn't. I guess that would make good copy, Leland, but the fact is I collect robots — I actually don't even test my robots or wind 'em up. I never did that with my toys either, like by Lehmann windups. I never knew if they worked unless somebody asked me. Mainly I collected 'em as sculpture. That's really how I got into folk art because I realized I was collecting

OME DOME, MEGO, Japan, 11″

LAUGHING ROBOT, S.H., Japan, 10″

LUNAR SPACEMAN, 12″

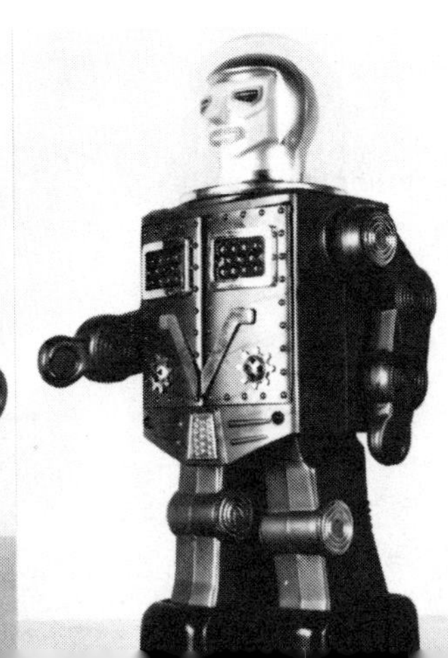

OUTER SPACE ROBOT, Hong Kong, 10″

ELECTRIC ROBOT, LEWIS MARX, U.S.A., 14″

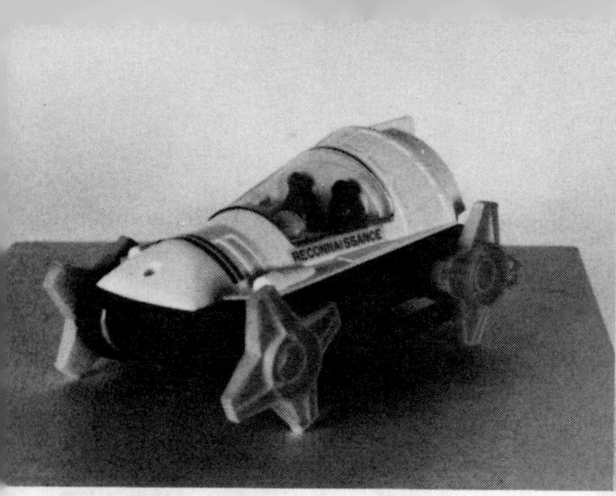

CA-STAR, ASHAKI, Japan

SPACE GUARD PILOT, ASAK, Japan, 8″

SPACE PACER, T.M., Japan, 7″

sculpture, not children's playthings. . . . Cast iron toys are like bronzes . . .

WHAT DO YOU THINK ABOUT AMERICAN SCIENCE FICTION TOYS AS COMPARED WITH THE JAPANESE TIN ONES?
You know, I think they're all fabulous. I don't really have a preference where they come from. And I don't have a preference whether they're metal or plastic. I just think it's the form — the more garish and outrageous, the more I like 'em.

WHAT ABOUT CONTEMPORARY, BRAND NEW TOYS?
Well, I make the rounds of the toy stores a couple of times a year and I don't know I probably buy about one out of five things I see. or one out of ten.

WHAT ABOUT STAR WARS?
I think STAR WARS is different than robots and space toys and I don't know that is but I think that's a whole different field, STAR WARS, and I think Trekkie stuff is a whole different field. Well, like the guy says "I'm a toy collector" but you won't find any banks in his collection or dolls. Doll collectors collect dolls, bank collectors collect banks and toy collectors collect things that are toys, that are out of proportion . . . Most banks are all in total proportion, they're like little miniature sculptures in real life.

IT LOOKS LIKE THEY'RE GOING TO KEEP THE STAR WARS STUFF IN PERMANENT PRODUCTION.
I think they made more STAR WARS things in the first round than any 10 robots, if you know what I mean. . . . Basically the stuff's (space toys) tooled 18 months in advance of its selling so they don't even known if it's going to be good. They've already stopped the run and they're tooling for another toy before they even find out if the other toy is successful.

ROBBY ROBOT, MORT TOYS, Hong Kong, 9″

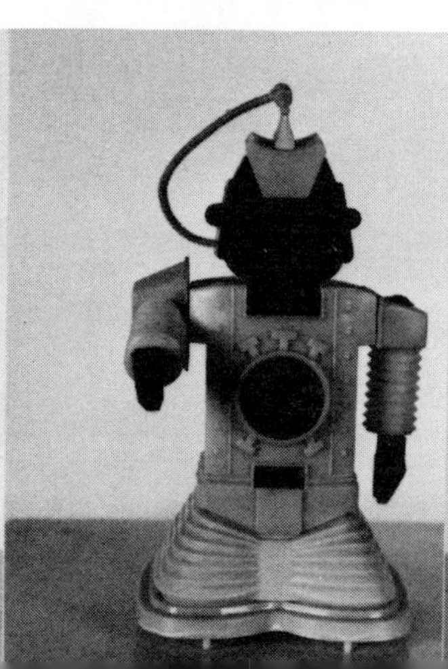

MAGNOR, CRAGSTAN, Japan, 9″

MYROBO, Japan, 9″

ROBOT 2500, DURHAM INDUSTRIES, Hong Kong, 10″

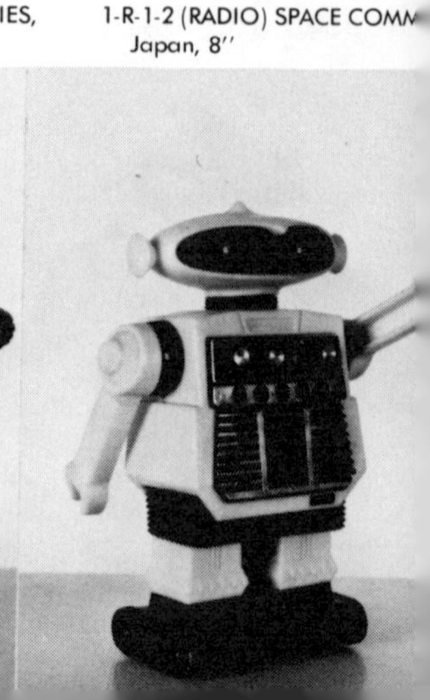

1-R-1-2 (RADIO) SPACE COMM Japan, 8″

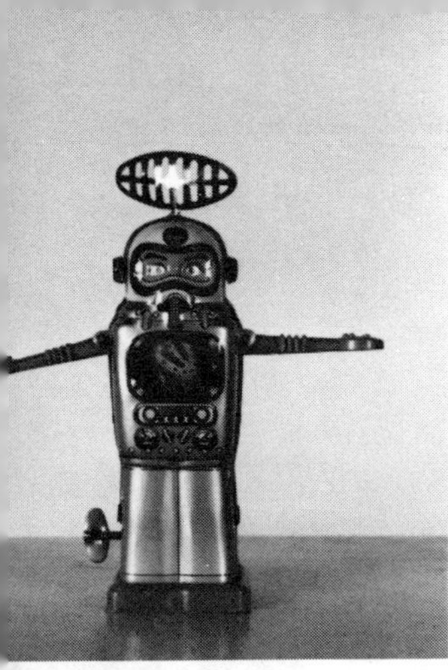

ECHANICAL TELEVISION SPACEMAN, ALPS, Japan, 7″

ROBOT (?), Greece, 6″

MECHANICAL SPACEMAN, S.Y., Japan, 6″

CAPTAIN ASTRO, Japan, 6″

DO YOU HAVE ANY ADVICE FOR PEOPLE BEGINNING TO COLLECT SPACE TOYS?
I think a person should buy the exact toy that appeals to them and that they like to look at, instead of collecting because they think it's gonna become famous or good because really no one can tell which ones are gonna be great. I don't care . . . You know I predict a lot of markets and I feel like if you pick something you really like it becomes good on your own taste. I've never bought a thing because I really thought it was gonna be valuable. That was in the back of my mind but the main reason I bought something was because I really, truly loved it. I think people should buy things they truly love and quit screwing around with thinking this stuff's all an investment. . . . They get burned out. They get a whole bunch of stuff they don't care about and they get bored with it. The commitment is that you're buying pieces of sculpture that are mechanical and move and have great color — you know there's not many things you can buy for $8 to $15 or $20 that'll really move you. Candy bars are gonna be $8 pretty soon.

HOW DID PEOPLE RESPOND TO THE SHOW PUT ON BY THE OTIS/PARSONS GALLERY?
Actually, we created a whole bunch of new collectors . . . the fact is they were in the arts. People constantly are becoming robot collectors coming to my gallery and I really don't even have a display of my robots out. I always have the pieces in transition that I buy and stick around on their way to the warehouse and . . . I give toys away constantly, robots, to friends' kids and to friends and I'll tell you I don't know how many collectors I've made but I'll bet it's at least a hundred just accidentially them seeing them around the gallery. The thing is they're in an environment that they look like art in here. People have to see them in that environment to realize that it's time to collect them, you know.

Photographed by Kate Loye

STAR S., TOMY TOY, Hong Kong, 8″

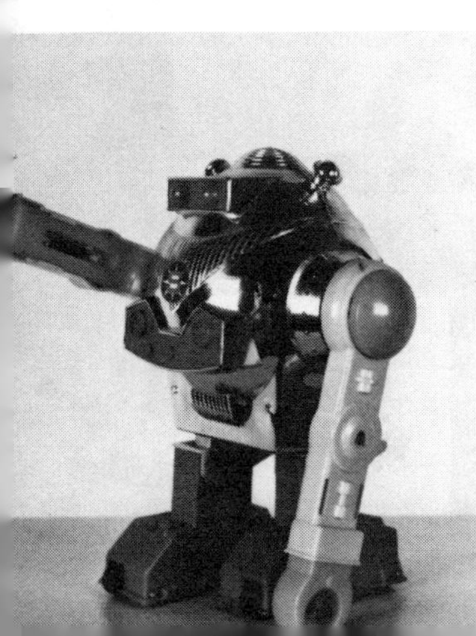

TR-2 ROBOTANK, TN, Japan, 5″

ROBOT TANK II, Japan, 10″

SKY ROBOT, S.H., Japan, 8″

"I've been thinking about space, too, since I was a child. I used to read BUCK ROGERS in the funny papers and listen to the radio program, and I always wondered if we'd have to wait until the 25th Century to get around out there the way he did."

1959, Gordon Cooper, One of the original 7 Mercury astronauts.

BIBLIOGRAPHY

Aldiss, Brian W. BILLION YEAR SPREE. London.

Ash, Brian, ed. THE VISUAL ENCYCLOPEDIA OF SCIENCE FICTION. New York, Harmony Books, 1977.

Boogaerts, Pierre. ROBOT. Paris, Futuropolis, 1978.

Brootin, Jacky, ed. LES JOUETS JAPONAIS. Avrigny, France, Anamorphose, 1982.

Buhler, Michael. TIN TOYS 1945-1975. New York, Quick Fox, 1978.

DeVincenzo, Ralph. FLEA MARKET SHOPPER. New York, Dafran House, 1972.

Gatland, Kenneth. THE ILLUSTRATED ENCYCLOPEDIA OF SPACE TECHNOLOGY, New York, Harmony Books, 1981.

Harman, Kenny, COMIC STRIP TOYS. Des Moines, Ia., Wallace-Homestead Book Co., 1975.

Hertz, Louis H. THE TOY COLLECTOR. New York, Hawthorne Books, 1977.

Herzog, David Allen. COLLECTING TODAY TOMORROW. New York, Arco Publishing Co., 1980.

Horn, Maurice, ed. THE WORLD ENCYCLOPEDIA OF COMICS. New York, Chelsea House Publishers, 1976.

Malone, Robert. THE ROBOT BOOK. New York, Push Pin Press/Harcourt Brace, Jovanovich, 1978.

Naha, Ed. THE SCIENCE FICTIONARY. Wideview Books, 1980.

O'Brien, Richard. COLLECTING TOYS. Florence, Ala., Books Americana, 1980.

Pressland, David. THE ART OF THE TIN TOY. New York, Crown Publishers, Inc. 1976.

Reichardt, Jasia. ROBOTS. New York, Penguin Books, 1978.

Robinson, Jerry. THE COMICS. New York, G.P. Putnam's Sons, 1974.

Sansweet, Stephen J. SCIENCE FICTION TOYS AND MODELS, VOL. 1. New York, Starlog Press, 1980.

Sugar, Bert Randolph. COLLECTIBLES: THE NOSTALGIA COLLECTORS BIBLE. New York, Quick Fox, 1981.

ACKNOWLEDGEMENTS

We wish to thank the following parties and to recognize the copyright owners: STAR WARS/THE EMPIRE STRIKES BACK © Lucasfilm, Ltd. and Twentieth Century-Fox Film Corp. KENNER PRODUCTS. LOST IN SPACE © CBS Television. STAR TREK © Paramount Pictures. BATTLESTAR GALACTICA © Universal City Studios. SHOGUN WARRIORS © Marvel Comics Group. MAGNUS ROBOT FIGHTER © Gold Key Comics. MY GREATEST ADVENTURE © D.C. Comics. THUNDER AGENTS © Tower Comics. FLASH GORDON © King Feature Syndicate. BUCK ROGERS © MCA TV. AMAZING STORIES © Street and Smith, © (renewed) The Conde Nast Pub. Inc. LIFE © TIME-LIFE, INC. NATIONAL AERONAUTICS AND SPACE ADMINISTRATION. WHITELEY-LOYE GALLERIES, Los Angeles, S.H. Toys, Japan. MAINE ANTIQUE DIGEST. We would also like to thank any other copyright holders who are mentioned in this book. If there are omissions or errors please contact the publishers and corrections will be made in future editions.

BULLETIN: ON A TROUBLED PULP PLANET AMONG DISTANT CELLULOID STARS THE THREAT TO MANKIND HAS BEEN TEMPORARILY HALTED......

BEAM US A MESSAGE SPACE TOY COLLECTORS! We are planning an expanded book with color illustrations on Space Toys and Robots. If you have toys, memorabilia or information you think should be included please drop us a line or card and let us know what you have. We will credit our collaborators and supply them with a free copy of the new book. We would welcome corrections, comments or suggestions on this or our next book project. **MAY YOUR D CELLS STAY CHARGED AND YOUR BATTERY BOX BE FREE FROM RUST!**

Copies of **SPACE TOYS** are available from the publisher at \$9.95 + \$1.00 postage. The **PRICE GUIDE** is \$3.50 ppd.

Dealers please write for discount.

We are always looking for SPACE TOYS and ROBOTS and SPACE MEMORABILIA to buy. Feel free to offer such items to us.